FORM & FACADE

contemporary facade

Spacetime

도서출판 시공문화사
120-796, 서울시 서대문구 현저동 200 극동 5층
등록 1993년 3월 12일
발행인 김기현

Spacetime Publisher
5F Kukdong bldg, #200 Hyunjeo-dong, Seodaemun-ku
Seoul, 120-796, Korea
Tel: +82 2 3147 1212, 2323
Fax: +82 2 3147 2626
www.spacetime.co.kr
spacetime@korea.com
Publisher_Kim Ki Hyun

architectuREpublic
http://architecturepublic.blogspot.com/
architecturepublic@gmail.com
creative director _sungminlee
editor _sungminlee, yeonjung kim, hyojin bae
translator _taewoong yoon, hyojin bae

design _typepage(www.typepage.com)

On a vast urban plan, MUSAC draws the scenario for art with the same optimistic attitude used by the Roman surveyors to sketch their cities in the landscape. In contrast to other types of spaces which, in their museum facet, focus on the exhibition of frozen historic collections, MUSAC is a living space that opens its doors to the wide-ranging manifestations of contemporary art. This is an art centre that constructs a set of chessboards on which the action is the protagonist of the space; a structure that develops from an open system, formed by fabric of squares and rhombi, and permitting the construction of a secret geography of memory.

MUSAC is a new space for culture, regarded as something that visualises the connections between man and nature.

A cluster of chained but independent rooms permit exhibitions of differing sizes and types. Each of the jaggedly shaped rooms constructs a continuous yet spatially differentiated area that opens onto the other rooms and courtyards, providing longitudinal, transversal and diagonal views. Five hundred prefab beams enclose a series of spaces that feature systematic repetition and formal expressiveness.

Outside, the public space takes on a concave shape to hold the activities and encounters, embraced by large coloured glass in homage to the city as the place for interpersonal relationships. Inside, a large area of continuous, different spaces, spattered with courtyards and large skylights, shapes an expressive system that speaks to us of the interest shared by architecture and art: the contemporary manifestation of the variable and the perennial, of equality and difference, of universality and transience, an echo of our own diversity and equality as people.

In its size, as a single storey building with white concrete walls and large coloured glazing seen from the outside, MUSAC strives to be a space where art is at ease and helps to erase the boundaries between private and public, between work and leisure and ultimately, between art and life.

Mansilla & Tuñón

MUSAC, 공동체의 얼굴

실험은 비문자적인 상황을 설명하는 어떤 텍스트이고, 다른 사람들이 그것이 단순한 텍스트인지 아닌지를 결정하기 위해 검토해 볼만한 텍스트이다. 최종 검사가 성공적으로 끝난다면 그것은 단순한 텍스트가 아니며 사실상 그 이면에는 실제 상황이 존재한다.

브루노 라투르 [1]

현재가 아직 구축되지 않은 동안 과거와 미래는 새로운 형태를 입는다. 매 순간마다 각각의 새로운 행위는 이미 행해진 것에 대한 수정을 강조하고, 집단 기억과 미래의 프로젝트 모두를 중단 없이 변경하며, 새롭거나 기발한-아마도 아직 알려지지 않은-윤곽을 제공한다. 변화하는 시나리오 안에서 불변하는 건축양식의 과거와 미래를 통해 개연성은 오직 가능한 확실성의 형태로 현실에 접근하도록 해 주는 오직 하나뿐인 얼굴로 전환된다. 이러한 변화의 중심에서 공간에 대한 정의를 단지 요구된 그 일-인간의 행위가 펼쳐지는 인공환경의 구축-의 작은 부분으로 여기며 건축은 그 범위를 더 넓게 확립하거나 그 동공을 팽창시킨다. 더 정확하게 말하면 공동체의 시간과 범주, 그 중에서도 후자는 브루노 라투르가 정의한 방식으로 이해된다.
"새롭게 떠오르는 패러다임에서 우리는 공동체-단체의 내부에서 일어나는 인적, 비인적 자산의 교류-라는 개념으로 그 많은 함축적 의미를 지닌 '사회'라는 단어를 바꾸어 놓았다." [2]
이것은 부분들 간의 협상이 기본적인 도구가 되는 일종의 전세계적 동원이고, 집단적 정체성을 격려하는 의지의 실제적인 혹은 가상의 시나리오를 창조하는 것을 목표로 한다. 여기에는 어떤 확장이 수반되는데 우리가 무엇인지 혹은 우리가 우리를 둘러싸고 있는 것과 동일한 수준에 있다고 생각하는 것, 그래서 정말 중요한 것은 인간, 자연, 기계, 혹은 가상의 것들과의 관계를 증대시키는 능력이고, 사람은 자기자신에 대해 실제적인 고찰을 할 수 있는 능력에 대해 감사를 표명하는 자연의 일부라는 것을 인식할 수 있는 공간을 만들어주기 위해 그것의 손상, 다툼, 합의, 그리고 거기 있는 차이점 내에서 세상에서의 존재감을 확장해 간다. 좀더 짧게 말하자면 우리는 끝없이, 지치지도 않고 돌아가는 세계의 한 부분 그 이상도 그 이하도 아닌 것이다.
오늘날 건축은 실제로 인간적이지 않은 것을 사회화하고 과학과 기술을 통한 인간들간의 관계가 형성된다. 건설되거나 상상되었던 인공물의 운명은 사용자의 손에 달려있음이 명백함에도 불구하고 작동중인 건축은, 브루노 라투르에 의하면 블랙박스를 열어 사실들과 기계들을 발명시기부터 설명하려고 하거나 인간과 비인간 사이의 연결에 대해 드러내려고 시도하는 것이다.
건축이라기 보다는 행위로 이해되는 MUSAC은 집합체의 순환시스템 구축을 시도하는 실험이다. MUSAC은 상호작용을 위한 연결과 결절점을 제공하는 새로운 공간이며, 그 결절점들은 인간과 자연, 기계, 인공물, 사실, 발생된 것, 혹은 상상된 것, 혹은 일어나려고 하는 가상의 것 사이의 연결을 시각적으로-혹은 가능한 것으로-표현하는 것이라고 이해할 수 있다.
거대하고 평평한 외피 위에 확장된 레옹시에서 호수에 가까운 규모인 MUSAC은 역사나 혹은 그로부터 남은 유물과 상호작용하고, 로마시대의 측량업자들이 풍경 위에 그 도시의 원천이 된 제7군단 진영의 윤곽을 계획했던 것과 동일한 낙천적인 방식으로 공연장의 경계를 결정했다.
포프의 비근대적인 비문맥화와 형태에 대한 두려움에 의해 생겨난 스칼라식 변환으로부터 물려받은 자치를 집행하는 수단에 의해 식민지 건설의 윤곽은, 건설될 기억의 비밀지형을 허용하는 광장과 장방형 광장을 조직함으로써 형성된 개방형 시스템의 발전된 구조인, 로마식 모자이크와 부합된다.
이것은 물결, 반사, 그리고 그것을 둘러싼 것과 상호작용하면서 변형되는 현상을 즉각 수용하는, 돌, 단어, 혹은 기억을 경계가 울퉁불퉁한 호수에 내던지는 것과 유사한, 우리를 둘러싼 것의 파편을 활성화시키거나 자극하는 과정과 관련된다.
이런 방식으로, 부지에 적용된 로마식 모자이크의 영향은 주변의 경작지를 분명하게 드러내면서, 생각지도 않은, 그리고 동시에 필연적인 경계를 활성화한다. 경작지는 중간부분에서는 질서가 잡혀있지만 주변에서는 무질서하고, 사막의 지평선처럼 고정되어 있으며, 두 가지 형상을 변경하는 것은 자연과의 부드러운 연합을 수반하는데, 이것은 카스티야와 레옹 안의 경작지 절반이 땅을 고갈시키지 않기 위해 매년 휴작지로 남을 것이 요구된다.
다른 곳을 살펴보면, 기하학과의 친밀함은 비카스티야식 구성의 잠재력을 향상시킨다. 실제로, 직교 구성은 원점을 지닌 각 지점간의 관계에 특권을 부여하는데, 이것은 위계적이거나 위계적이지 않은 질서를 구축하지만, 가까운 것과의 특별한 관계 즉, 어떤 특정한 것 또는 지속적인 것을 포함한 구별된 종류의 관계를 구축하기 위한 심각한 어려움에 빠진다. 그럼에도 불구하고, 씨실-날실 같은 조합은 중심을 참고하지는 않았지만, 대신 더 앞에 일어날 일을 꼭 알아야 할 필요가 없는, 가깝거나 연속된 것에 특권을 주고 상호작용하는 메카니즘처럼 행동한다. 지역적 행동 유형의 체계는 이런 식으로 구축되는데, 이것은 전체의 통일성이 완전한 형태의 분할이나 형성하는 구성요소에 의해 규정되지 않고, 그것들을 연결하는 관계에 의해 규정되는 구성요소 대 구성요소간의 관계를 생성해낸다. 그 물결은 코르도바 모스크인 바르셀로나의 라스 아타라자나스, 혹은 브뤼셀 전시회 때의 스페인 파빌리온의 영향력과 기억 아래, 여기 수리적 영역에서 반향을 일으킨다.
그 과정은, 진정으로 관련 있는 것들은 그것들의 연결, 분절, 변형 등이라는 추정하에, 내부적인 설명(내용 지향적인)과 외부적인 설명(맥락을 향한 그것들의 시각을 관리하는) 사이의 차이점을 극복하기 위해 애쓰는 움직임과 유사하다.
건축의 배경을 가까운 것에만 주의를 기울이는 행동계획과 융합함으로써(정방형 광장은 바로 옆에 장방형이 있다는 것을 아는 경우가 거의 없다), 우리는 주제를 반복하지만, 어디서건 잘려나가서 그 부분 없이 거기서부터 그 상태로 앞으로 나아갈 수 있는 직물

구조와 유사한 형상 앞에 놓인 스스로를 발견한다. 이것은 마치 새들의 한 무리는 몇 마리의 새를 더하거나 뺀 것 그 이상도 그 이하도 아닌 것과 같다. 이것은 중요한데, 우리는 더 많은 조각이 더해지거나 감해지면, 혹은 심지어 이것들의 크기가 줄어들거나 늘어나도, 변하지 않고 남아있는 특징을 가진 행동계획에 도달하게 되기 때문이다.
건물의 경계나 형태는 전혀 중요하지 않게 되고, 건물의 구획은 요구된 실제 크기에 맞추어질 때까지 동등한 타당성을 가진다. 각 구획은 대등하거나, 가능하다면 동등한 것이 되고, 우리의 마지막 기억은 변하지 않고 남게 될 것이다. 그렇지만, 여기서 타당한 것은 우리의 기억이 아니고, 가능성과 그 너머의 자유에 관한 개념의 형상이다.
동일한 것과 다른 것의 존재 내의 자유는 주변의 상황을 통해서만 최종적으로 드러나는 구체적인 형태를 야기하면서, 작업을 충만하게 만드는 반영의 영역으로 전환된다. 반영의 공동 영역에 대한 각 사업, 각 행위, 각 견해의 연결은 그 자체로 형태와 무관한 추상적인 요소를 부여 받은 작업을 가능하게 한다. 어떤 심상이나 흥미들의 충동은 이제 구체적인 상황에 따라 특정 지워진다.
따라서, MUSAC의 인테리어는 연속적이면서도 구별되는 공간적 이벤트가 연달아 구성되었는데, 여기에는 파티오와 거대한 천창들이 군데군데 있어서, 인간인 우리의 다양성과 동등성에 대한 반향으로서의 차이점과 동일한 점, 보편적인 것과 일시적인 것의 현대적 표명인, 건축과 예술이 함께 공유하는 관심에 대해 말하는 표현 체계에 형태를 부여한다.
이런 약간 무의식적인 방식으로, MUSAC는 명확한 윤곽을 얻게 되는데, 그것은 포기할 수 없는 인간 조건의 물질적인 반향으로서, 일련의 정확한 규칙, 질서와 자유가 동시에 존재하는 게임 보드, 차이점과 공통점이 존재하는 다른 면 등이다.
미술관의 우수성이 고정된 역사적 전시물들을 보여주는 것에 집중되어 있는 다른 전시공간과는 달리, MUSAC는 광범위한 예술적 표현양식에 문호를 개방한 살아있는 공간이다. 행위가 공간을 이끌어가는 주체가 되는 일련의 게임보드를 만드는 예술센터이고, 각기 다른 크기와 특징을 가진 전시회가 만들어지는 자치적이고 상호 연관된 일련의 전시실들이며, 비정형적인 형상의 각 전시실은, 공간적으로는 분리되었지만, 다른 전시실과 파티오에 종, 횡, 사선 방향으로의 장면을 제공하는 연속적인 공간이다.
이것은 기하학적으로, 또한 구성적으로 유사한 일련의 전시실들이지만, 천창, 파티오, 회화창을 통해 들어오는 빛의 종류가 각각 다르기 때문에, 사실적인 것과 인공적인 것의 독립화 및 연계과정의 동시적 과정에 의해 다른 성질을 가진 공간으로 변형된다. 500개의 미리 제조된 보들은 변화하는 빛이 유도하는 체계적인 반복과 형식적인 표현으로 특징 지워진 일련의 공간들을 완결 짓는다.
이런 동일성과 차이점이라는 상황의 존재함으로서, 두 개의 커다란 천창이 빛을 공간으로 공급하는 출입구 로비에 의해 확장되지만, 그것들 각각은 다른 방향으로부터 빛을 흡수하는데, 하나는 동쪽에서, 다른 하나는 서쪽으로부터 흡수한다. 그럼으로써, 아침에는 그 중 하나가 따뜻하고 무거운 빛을 떨어뜨리고, 다른 하나는 차갑고 변화하는 빛을 끌어들인다. 오후가 되어 태양이 원을 그리면 상황은 반대가 되는데, 마치 두 개의 천창이(혹은 두 사람이) 서로의 다양성을 인식할 수 있을지도 모른다. 그러나 하루가 끝날 즈음에는 단일한 조직의 일부로서 서로 바꿀 수 있는 것으로 상대방을 인식하게 된다.
외부에서, 공공공간은 개인적인 상호연락장소로서 도시에 대한 경의를 표하면서, 거대한 색채 유리판에 의해 둘러싸인 형상인, 서로 다른 행위와 만남을 수용하기 위한 오목한 형상을 드러낸다. 구조물의 잘라낸 부분은 부정형의 경계를 형성하기 위해 사용되는데, 그것은 헌신적인 조사의 결과라기 보다는 뜻밖의 습득물에 더 가깝다.
이 포럼 벽체의 표면은 공공적인 이미지를 만들어내는 주역으로서, 그것을 위한 설정과 관련된 색채들로 장식되었다. 그것들은 건조세척에 대해, 그리고 호기심 많은 어린이들이 창 밖으로 고개를 내밀고 바라다볼 도시의 광장에 있는 발코니로부터 매달린 나무의 생명력에 대해 설명하고 있다. 처벌받지 않은 약탈의 역사에도 불구하고, 그것의 이미지는 레옹 성당의 스테인드 글라스 조각의 픽셀화로부터 유래된다. 이것은 추상화의 과정, 혹은 결여된 것, 혹은 엘 할코네로의 형상에 대한 냉소적인 시각이다. 형태와 동떨어진 팰코너는 송골매가 그 자신으로부터 멀어지는 만큼 거리를 갖지만, 의심할 바 없이 개인적인 거리를 두고, 그것의 질서와 무질서의 통합, 그 자체의 복잡함을 지닌 자연만이 만들어낼 수 있는 겹침과 분리가 통합된 것과 관계를 맺는다. 그것은 우리 시대의 안목 즉, 컴퓨터를 통해 가장 오래된 것을 바라보는 것과 관련된 문제이다. 이런 것들만이 인간의 작품들이 자연의 조각처럼, 혹은 미래의 인류학적 자취로 여겨진다는 면에서 자연에 가까이 다가가도록 한다.
실제로 우리는 현실을 구성하지도 않고, 현실이 우리를 구축하지도 않지만, MUSAC는 오로지 공공과 사적 영역의 경계, 여가와 업무, 그리고 최종적으로는 예술과 인생의 경계를 허물어뜨리기 위한 방안을 찾고 있고, 그로 인해 그것의 지붕은 카스티야와 레옹의 모든 지방을 관통하는 듀어로 강을 통합한다. 마치 예술과 강물이 그것들이 영원히 동일해 보이고 영원히 새로워 보이도록 만드는 그런 지속적인 순환고리를 공유할 수도 있는 것처럼, 혹은 아마도, 그것들이 단순하고 평범하게 매일 흘러가는 강물을 실어 나르는 본연의 임무를 다하는 것이다. 그저 강물이 그러하듯이.

만실라 튜논 건축사무소

MUSAC, THE FACE OF THE COLLECTIVE

An experiment is a text that narrates a non-textual situation, a text that others will later supervise in order to decide whether it's a simple text or not. If the final test meets with success, then it isn't a mere text, there's a real situation behind it, in fact.

Bruno Latour[1]

While the present is yet being constructed, the past and future assume new forms. Each instant, each new action, highlights a revision of what is done and also gives a new, or novel—perhaps still unknown—outline to that which is to be done, uninterruptedly modifying both collective memory and future projects.

In this changing scenario, with past and future in constant construction, probability turns into the only possible appearance of certainty, into the only face that enables it to approach reality. At the heart of this transformation, architecture fixes its gaze more widely, or dilates its pupils, considering the definition of space as only a small part of the task to which it is summoned: the construction of artificial environments in which the acts of men unfold. Or to put it more precisely, the time and the territory of the collective, the latter understood in the way Bruno Latour defines it: "In the new emerging paradigm, we have replaced by the notion of a collective—defined as the exchange of human and nonhuman properties in the bosom of a corporation—the word 'society', so burdened by its many connotations." [2]

This is a sort of mobilisation of the world in which the basic tool is negotiation between the parts, and the goal the creation of real or virtual scenarios of wills that encourage collective identity. An extension is involved, putting what we are or think we are on the same level as that which surrounds us, so that the truly important thing is the ability to multiply the relations between humans, nature, machines or the virtual, so that in its abrasions, in its conflicts, in its agreements, and in its distances there dilates our being in the world so as to make space for the awareness that man and woman are nature unfolding thanks to its ability to think about itself virtually. Or put more rapidly: that we are nothing less, but also nothing more, than a small part of a world that spins endlessly, tirelessly.

Today, architecture socialises that which is not really human, so that there may be established, through science and technology, relations with humans. And although it is clear that the destiny of the artefacts constructed or imagined is in the hands of the subsequent users, architecture in action attempts, going back to Bruno Latour, to open the black boxes, explaining facts and machines from the discovery or the revelation of the connection between humans and nonhumans.

The MUSAC, understood not so much as architecture but as action, is an experiment which attempts to construct a circulatory system of the collective, a new space of links and nodes for interaction, the latter understood as that which renders visible—or rather possible—the links between humans and nature, machines, artefacts, the facts, occurred or imagined, or on the point of happening, the virtual.

On a huge flat surface, almost a lake in the expansion of the city of León, the MUSAC interacts with history, or what remains of it, manipulating it by defining the art venue in the same optimistic way that Roman surveyors laid out the lineaments of the Seventh Legion encampment—the origin of the city—on the landscape.

By means of a process of autonomisation, inherited from the non-modern decontextualisations of Pop and from scalar transformations born of the fear of form, the lineaments of colonisation correspond to a Roman mosaic: a structure that is developed from an open system formed by a fabric of squares and rhomboids which permit a secret geography of memory to be constructed.

This involves a process of activation or excitation of a fragment of what surrounds us, similar to casting a stone, or a word, or a memory into a lake with an uneven perimeter, which immediately receives the ripples, reflecting and deforming them when interacting with what surrounds it.

In this way the echo of the Roman mosaic cast into the location activates unsuspected, and at the same time logical, perimeters, distinctly illuminating the cultivated fields nearby, fields always ordered in the middle but disordered at the perimeter, immobile like the desert horizon, and the alternation of two figures convokes that gentle alliance with nature which calls for half the fields in Castile and León to be left to lie fallow each year in order not to exhaust her.

If we look elsewhere, familiarity with geometry updates the potentiality of non-Cartesian configurations. Actually, the orthogonal organisation privileges the relations of each point with an origin, establishing an order that may or may not be hierarchical, but runs into serious difficulties in order to establish particular relations with the close-at-hand, relations of a specific or differentiating kind with the contiguous.

Nevertheless, the warp and weft-like combination makes no reference to a centre, but behaves instead like a mechanism which privileges and interacts with what is close-at-hand or contiguous, without the need to know what is happening further on. A system of local pattern behaviour is thus established, one which generates the component to component relations, in which the coherence of the whole is not ordained by the division of a complete figure or by the components making it up but by the connection that links them. The waves rebound here on the terrain of mathematical fields, beneath the shadow and the recollection of the Cordoba mosque, Las Atarazanas in Barcelona or the Spanish Pavilion at the Brussels Exhibition.

The process is akin to a dynamic which tries to overcome the differences between internalist explanations(which look towards content) and externalist ones(which direct their gaze towards context) in assuming that what is truly relevant are their links, articulations and transformations.

By grafting the setting of the architecture with a scheme of behaviour that

only pays attention to the close-at-hand—the square hardly knows that it has a rhomboid next to it—we find ourselves before a figure similar to a woven fabric, which repeats a motif, but which can be cut away at any point without it foregoing its condition thereby, just as a flock of birds is no more or no less of a flock if a few birds are subtracted or added to it. This is important, because we arrive at a scheme of behaviour whose character remains unaltered if more pieces are added or taken away, or even if these are reduced or increased in size. The perimeter or shape of the building literally ceases to be of any importance, and any layout, until being adjusted to the real size required, is equally valid. Each layout is equivalent, or what amounts to the same thing, possible, and our final recollection will remain unaltered. Although here the relevant thing is not our recollections, but the appearance of the concept of possibility and, beyond that, of freedom.

A freedom in which the presence of the same and the different is converted into a terrain of reflection that inundates the work, acquiring concrete form only through the edge condition that ultimately appears. The connection of the project, of each action, of each thought, to a common territory of reflection enables the work to be endowed with a component that is abstract and, as such, independent of the form. A reflection or a vector of interests which is then particularised according to the concrete conditions.

Thus, the interior of the MUSAC is constructed as a succession of continuous but distinct spatial events, dotted with patios and huge skylights, giving shape to an expressive system that speaks to us of the interest that architecture and art share: the contemporary manifestation of the different and the same, of the universal and the transitory, as an echo of our own diversity and equality as people.

In this slightly involuntary way the MUSAC acquires a sharp outline: that of a set of precise rules, a game board on which order and freedom are simultaneously present, the other face of that presence of the different and the same, as a material echo of our unrenounceable human condition.

Unlike other kinds of space whose museistic quality is centred on the showing of fixed historical collections, the MUSAC is a living space which opens the door to a wide range of contemporary artistic expressions; an art centre that constructs a set of game boards in which actions are the leading players of the space itself; a series of autonomous and interconnected exhibition rooms enables shows of different sizes and characteristics to be created; each irregularly shaped room constructs a continuous, yet spatially separate, space that gives onto the other rooms and patios, providing longitudinal, transversal and diagonal views.

This is a set of rooms alike in their geometry and construction, but which, due to the different kinds of light coming from skylights, patios and picture windows, are nevertheless transformed into spaces with different qualities by means of a simultaneous process of autonomisation and alliance of facts and artefacts. Five hundred prefabricated beams close off a series of spaces characterised by the systematic repetition and the formal expressiveness a changing light brings.

This presence of the condition of sameness and difference is amplified in the entrance lobby where two huge twin skylights disburse light on the space, but each of them is receiving light from different directions, one east and the other west. And so, in the morning one lets a warm, heavy light fall, while the other lets in a cold, changeable light. When the afternoon descends, and the sun shifts round, the situation is reversed, as if two skylights, or two people, might be able to recognise their mutual diversity, but at the end of the day recognise each other as interchangeable, or part of a single group.

Outside, public space assumes a concave shape in order to accommodate different activities and encounters, a shape enfolded by great coloured panes of glass, in which homage is paid to the city as a place of personal intercommunication. Offcuts of the fabric go to form its irregular perimeter, and they seem more like windfalls than the outcome of a dedicated search.

The wall surfaces of this Forum are bedecked in colours, colours engaged in setting themselves up as protagonists of the Public Image; they speak to us of the vitality of the drying washing and the plants which hang from the balconies of the city squares, of inquisitive children leaning out of the window to look. Although ransacking history with impunity once more, its image comes from the pixelisation of a fragment of the stained glass windows of León Cathedral. This process of abstraction, or of an absent or cold gaze upon the figure of El Halconero, The Falconer, distances it from the form, distances it as much as a falcon distances itself, but refers without a doubt to that combination—at a distance from the personal—of order and disorder, of imbrication and detachment that only nature, with its complexity, is capable of producing. It is a question of seeing the most ancient through the pupils of our own era—computers. Only these manage to get close to nature, such that the works of man seem like a fragment of nature, or an anthropological vestige of the future.

Although in actual fact we neither construct a reality, nor does reality construct us; the MUSAC seeks only to efface the boundaries between the public and the private, between leisure and work, and, once and for all, between art and life, and so its roofs convoke the River Duero, which traverses all the provinces of Castile and León, as if art and water might share that continuous cycle which makes them seem forever the same and forever new—or, perhaps, they simply serve for what they are, for carrying along a simple and normal, everyday water. Just water.

Mansilla & Tuñón

NOTES

[1] LATOUR, Bruno, La esperanza de Pandora, ensayos sobre la realidad de los estudios de la ciencia, Gedisa, Barcelona, 2001, p. 149. (English edition: Pandora's Hope. Essays on the Reality of Science Studies, Harvard University Press, Cambridge, Mass., 1999).

[2] Ibid., p. 231.

\ the pixelisation of a fragment of the stained glass windows of León Cathedral

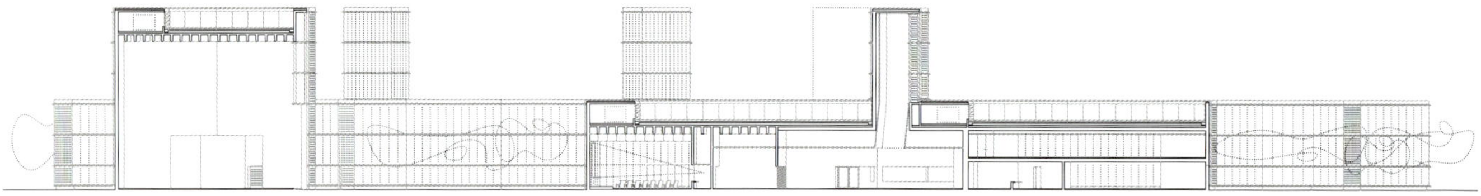

\ section 1

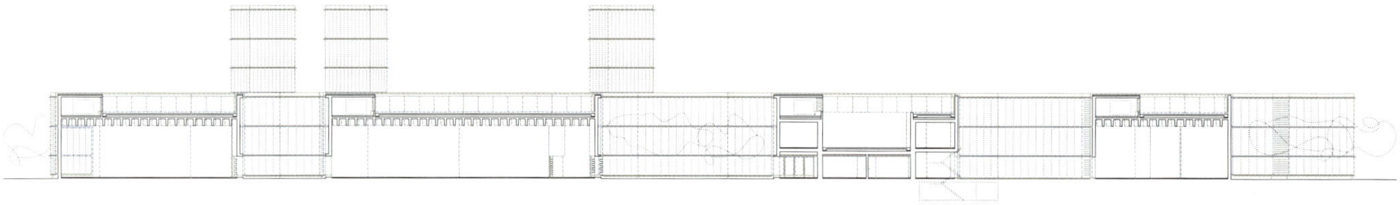

\ section 2

opening of MUSAC /

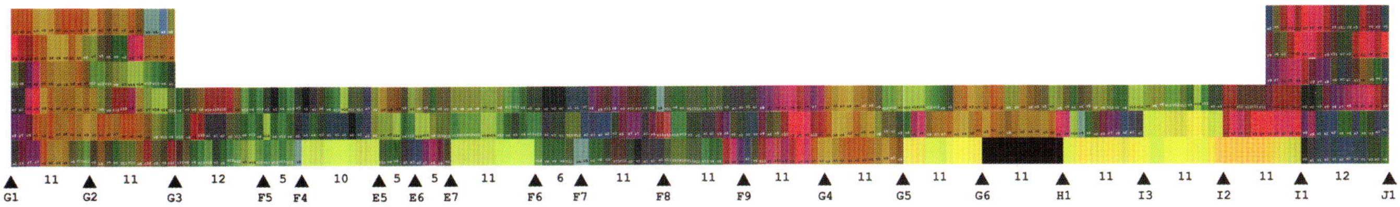

\ colors of facade

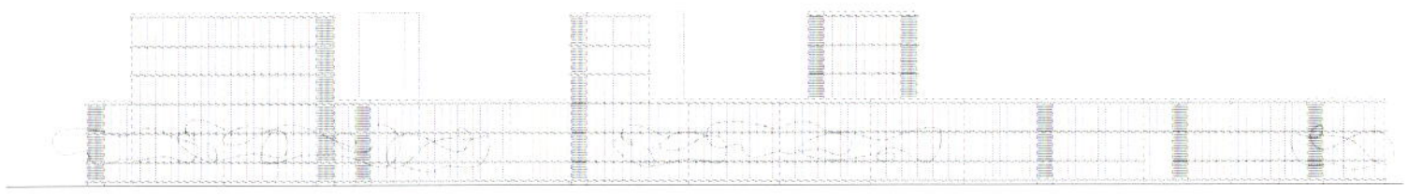

elevation 1 /

elevation 2 /

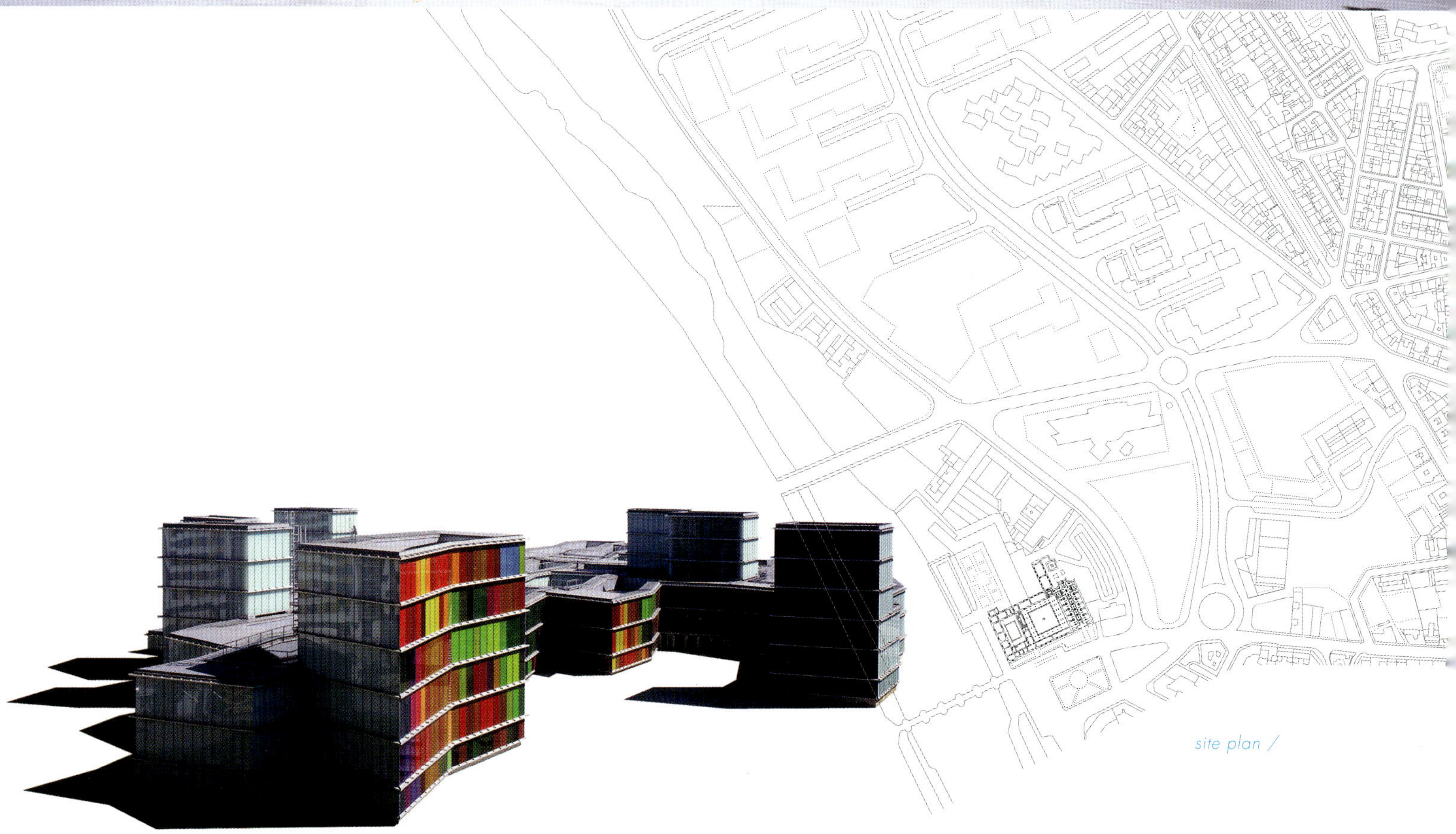

site plan /

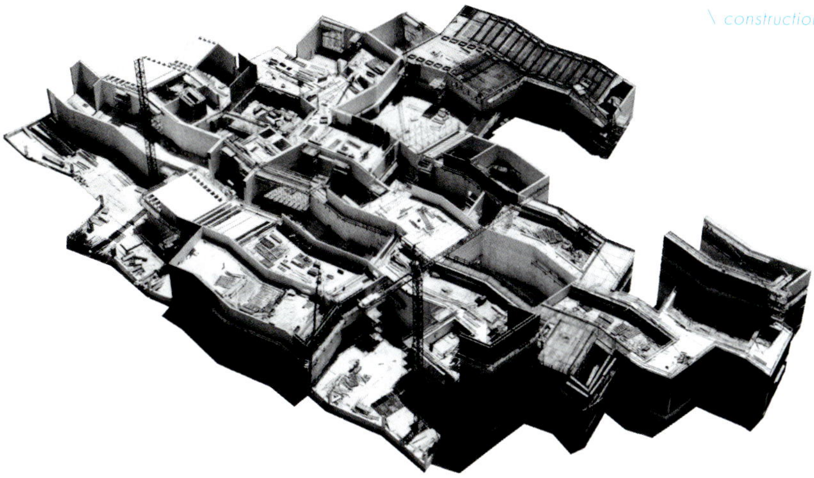

\ construction area

diagrams of hiperpunto /

\ ground floor plan + 1st floor plan

1.MAIN LOBBY
2.STAFF
3.OFFICE
4.EDUCATIONAL WORKSHOP
5 LIBRARY
6.TOILETS
7.SHOP
8.CAFE
9.MULTI_ USE ROOM
10.PATIO
11.RESTORATION WORKSHOP
12.STORAGE
13.TECHNICAL AREA
14.RESTAURANT
15.LOAD AND STORAGE
16.LOADING AREA
17.EXHIBITION ROOMS
18.EXHIBITION PATIO
19.PUBLIC COURTYARD

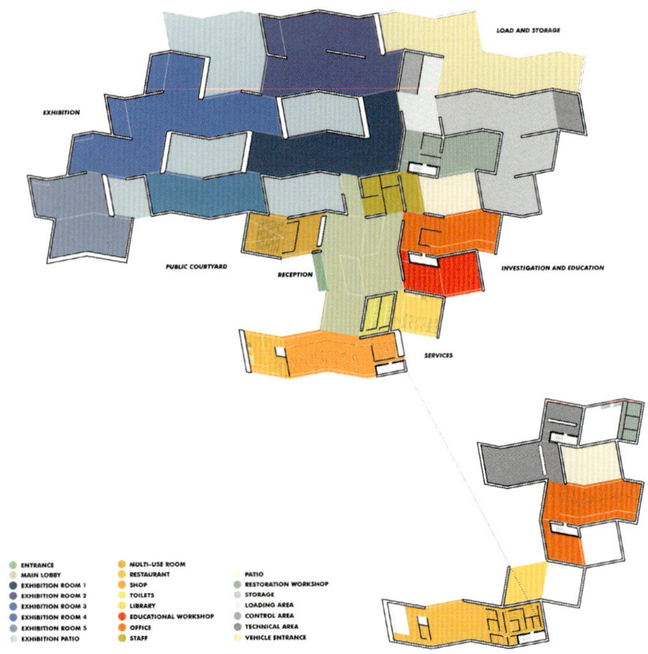

\ *function chart*

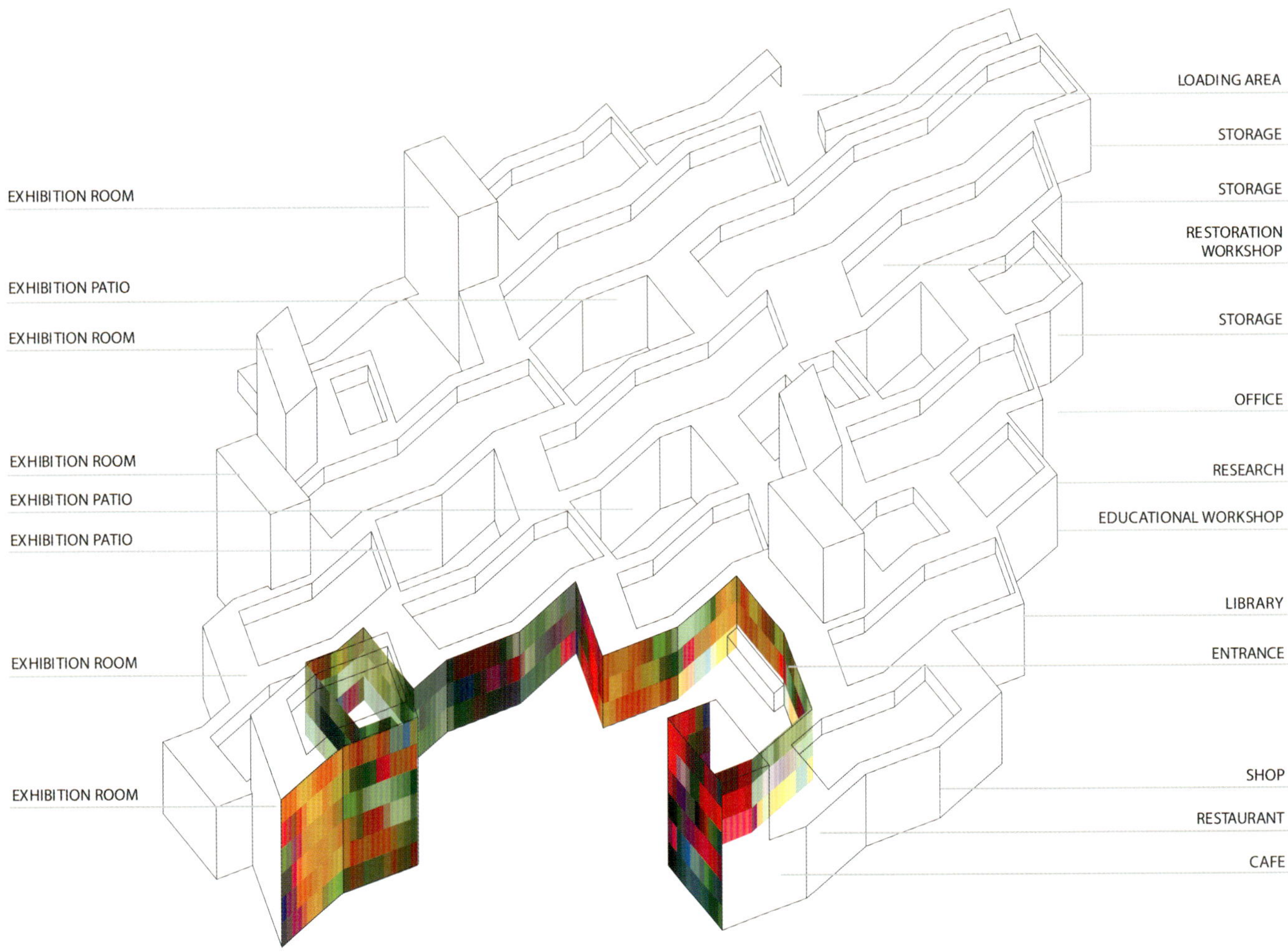

\ *color axono*

colors axon of construction /

\ colors montage

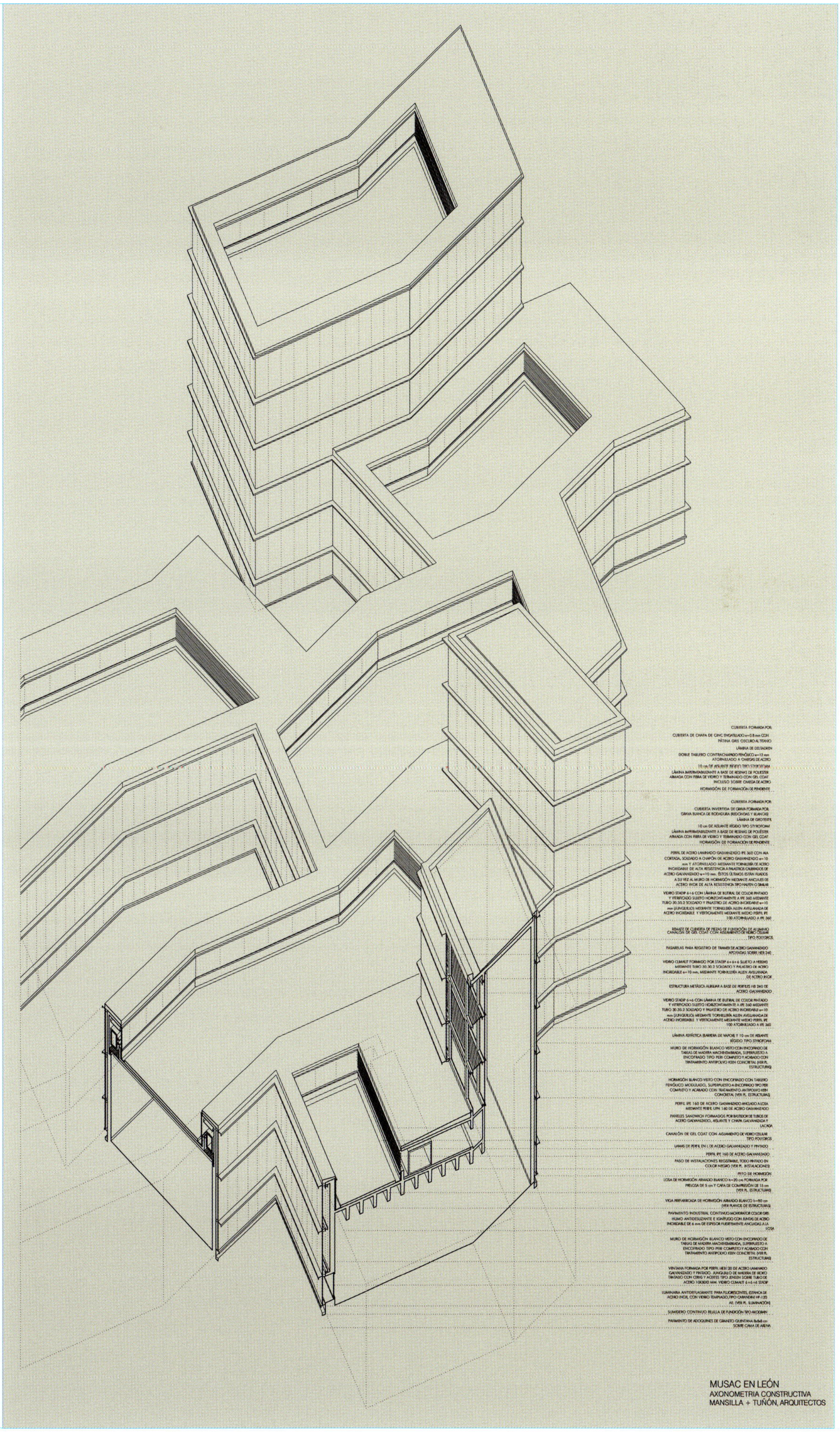
MUSAC EN LEÓN
AXONOMETRIA CONSTRUCTIVA
MANSILLA + TUÑÓN, ARQUITECTOS

Almere, the Netherlands

1999-2004

UN Studio

Function_ Office

Site area_ 8.833m^2

Building area_ offices(23,000m^2), parking(15,000m^2)

Photographer_ Christian Richters

Client_ Eurocommerce, Deventer

Project architect_ Ben van Berkel, Marco Hemmerling, Martin Kuitert, Henri Snel, Gianni Cito, Olaf Gisper, Yuri Werner, Marco van Helden, Eric Kauffman, Katrin Meyer, Tanja Koch, Igor Kebel, Marcel Buis, Ron Roos, Boudewijn Rosman, Stella Vesselinova

OFFICES 'LA DEFENSE'

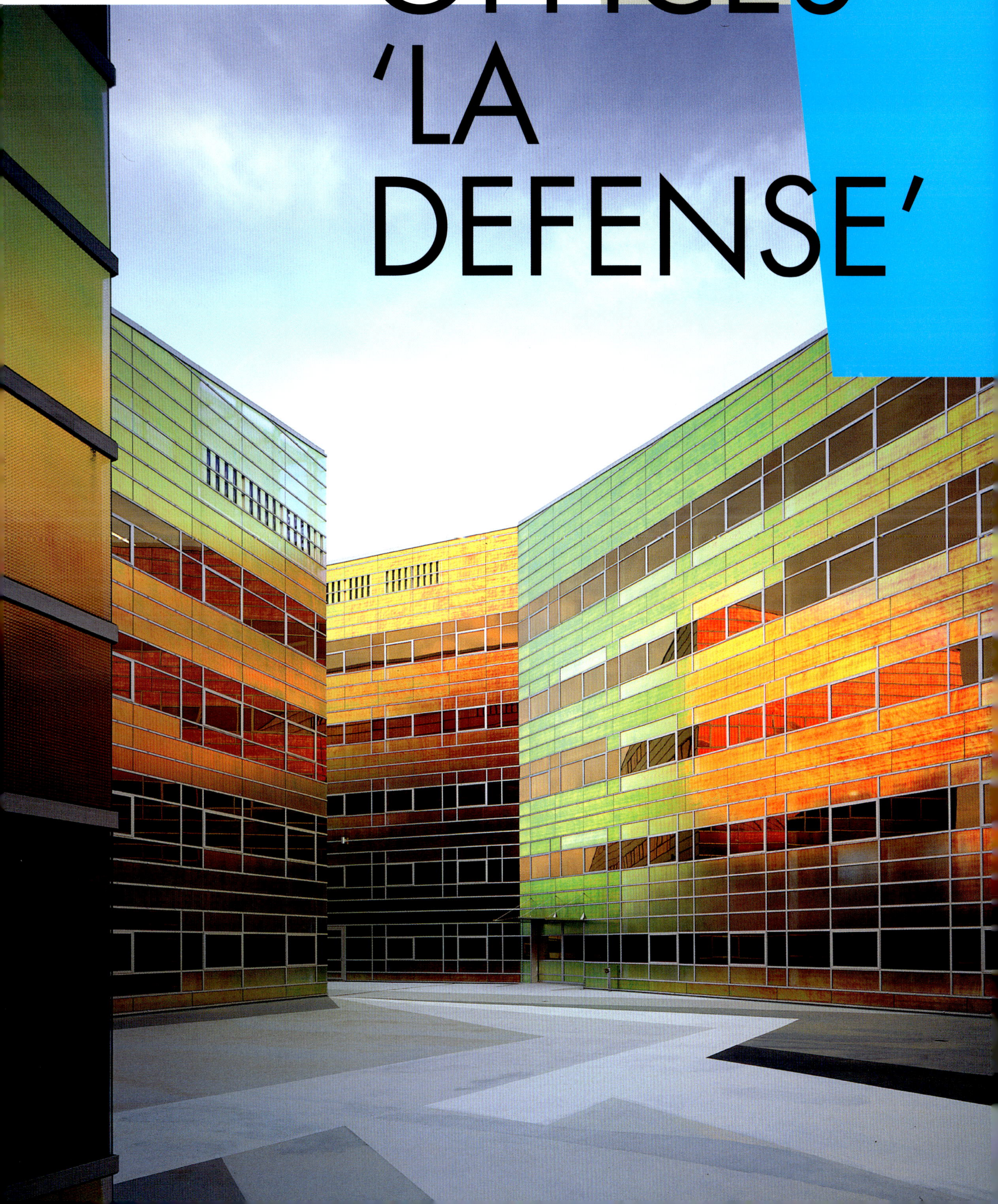

사무실 용도의 건물 '라 데팡스'는 금속으로 마감된 파사드에 주변 환경을 그대로 반영한 채 도시 맥락 속에 자신을 드러내며 서 있다. 바깥쪽의 표피는 도시성과 유닛의 막힌 정도를 나타낸다. 내부 중정으로 들어서면 건물이 본래의 특징을 드러내는데, 중정에 면한 파사드는 여러 가지 색채가 입혀진 박면이 삽입된 유리 판넬로 만들어졌다. 일과 시간 동안 파사드는 시간과 각도에 따라 노랑, 파랑, 빨강 혹은 보라, 녹색등으로 변화하기를 반복한다.

라 데팡스는 알메르(Almere) 시 중앙역에 자리잡고 있다. 평면을 보면 부정형의 대지를 차지하는, 길이와 높이가 다른 네 개로 구별되는 띠들이 적절하고 합리적으로 구성되어 있음을 알 수 있다. 설계안은 이러한 띠들을 알메레 타운의 대규모의 중심부 신규 개발과 연계하는 한층 통합된 도심개발계획과 일정 범위에서 연결한다.

근본적으로는 닫힌 이 건물은 공원과 건물 후면과 연결되는 두 지점에서 관통된다. 도심으로 지층 레벨을 들어 올림으로써 이 연결은 더욱 강조된다. 주차 시설을 유닛의 사이보다 지하에 배치함으로써 쾌적한 내부공간을 만들어낸다. 유닛들은 5에서 6, 또는 3에서 4 등의 레벨로 변화하고, 최상층은 2개 층으로 구성되기도 한다.

설계안은 많은 기업이 사용하는 콤플렉스를 기초로 한 것이다. 밀도와 프로그램은 단위면적당 가격을 낮게 유지하려는 방안으로 조정되었는데, 이러한 내용은 개발 사업이 지역 사람들에게 더 매력적으로 다가갈 것이다. 출입구의 수가 많은 이유는 건물의 접근성을 강조하기 위해서이다. 도로 쪽에서 유닛으로의 접근은 지층 레벨을 통해 이루어지고, 내부에서는 높여진 출입구를 통해 접근한다. 이런 구성은 사무실 블록의 유연성을 강조한다.

This office building 'La Defense' appears in its urban context as a modest building volume reflecting the direct surroundings in its metallic facade finishing. The outer skin expresses urbanity and a degree of closeness of the units. Entering the inner-court yard the building reveals its genuine characteristics. The facade adjacent to the courtyard is built-up out of glass panels in which a multi-colored foil is integrated. Dependent on the time of the day and the angel of incidence, the facade changes from yellow to blue, to red or from purple to green and back again.

La Defense is located behind Almere Central Station. The plan shows a neutral, rational organization with four separate bands of different lengths and heights covering the irregular plot. The proposal links these bands to some extent in a more integrated urban solution which ties in with the new, larger-scale center development of Almere Town.

The essentially closed complex is interrupted at two points, creating a link with the park to the rear. Raising the ground level from the town side strengthens the link. Parking facilities are provided under the building rather than in between the units, creating a pleasant inner area. The units vary in height from 5 or 6 to 3 or 4 levels, the top level often being doubled.

The design is based on the complex being used by a number of companies. The density and program are balanced in such a way as to keep the price per square meter low, making the development more attractive to local users. A large number of entrances are provided to highlight the complex's accessibility. On the street side, access to the units is via entrances at ground floor level, on the inside via raised entrances. The organization accentuates the flexibility of the office block.

UN Studio

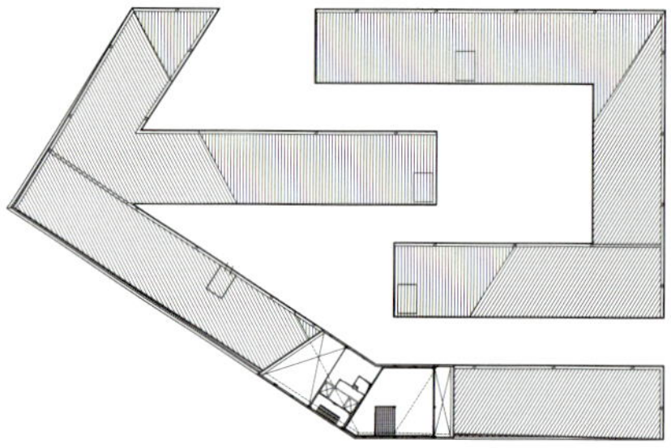

\ 6th floor plan

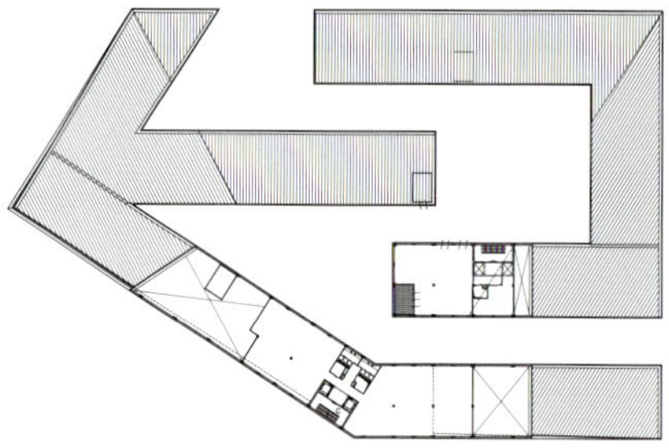

\ 5th floor plan

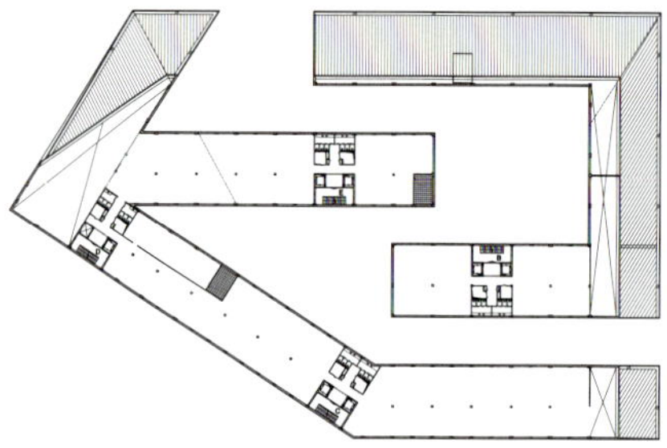

\ 4th floor plan

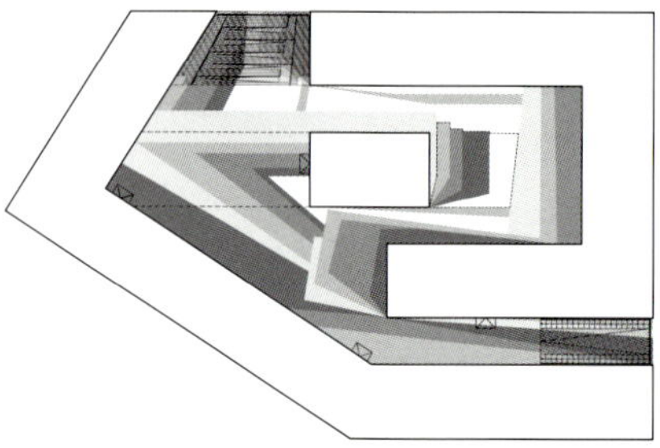

\ terrace

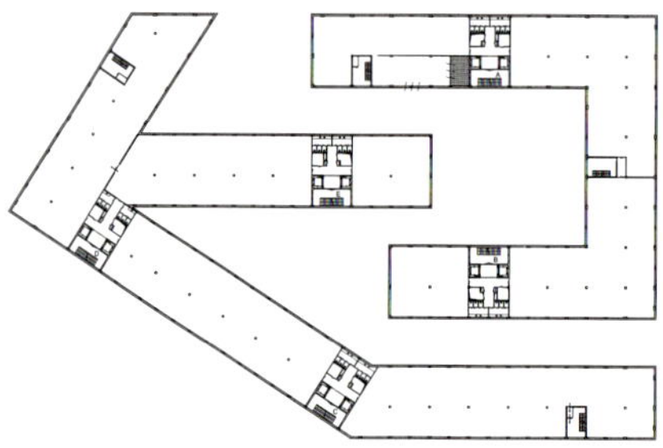

\ 3rd floor plan

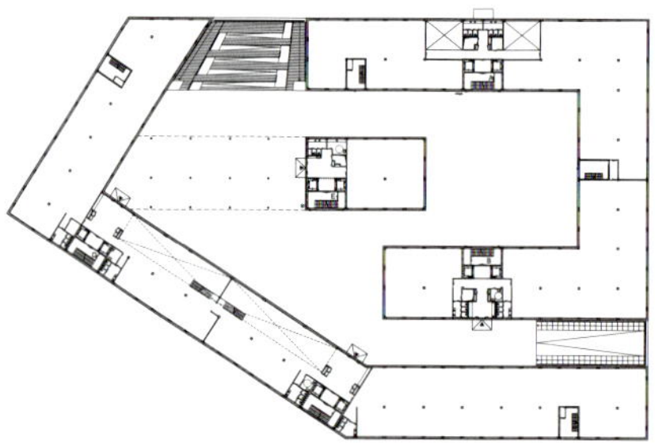

\ 1st floor plan

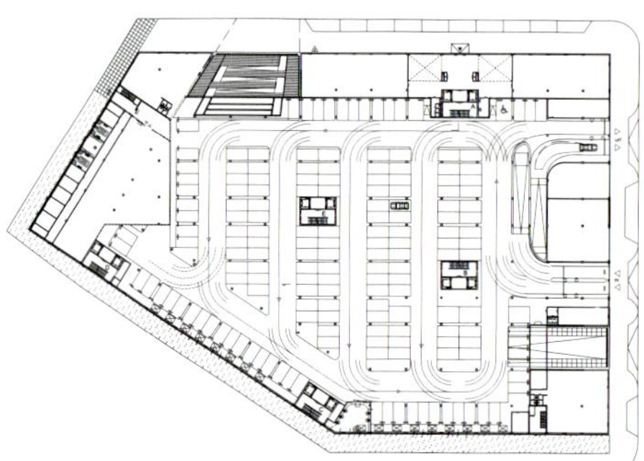

\ ground floor plan

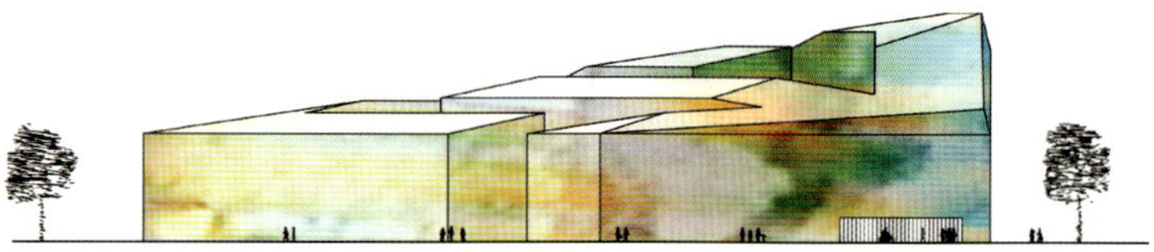

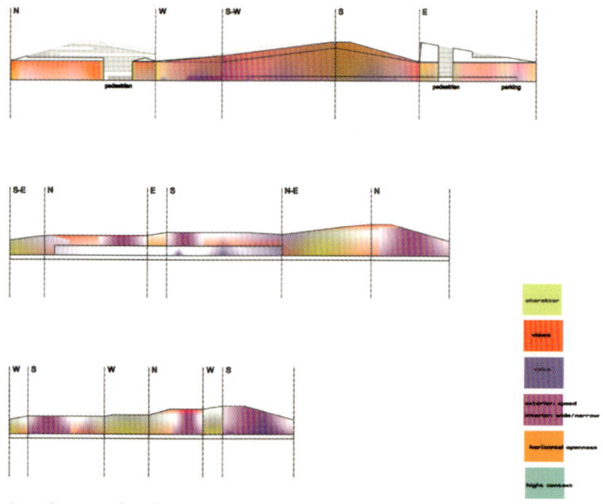

\ urbanscale elevation

\ projection of communication value

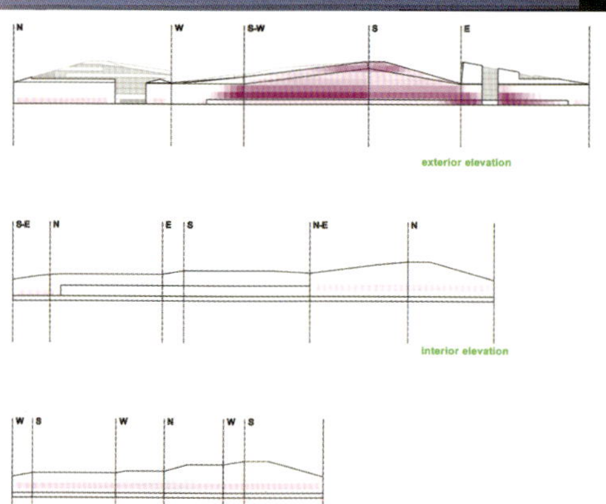

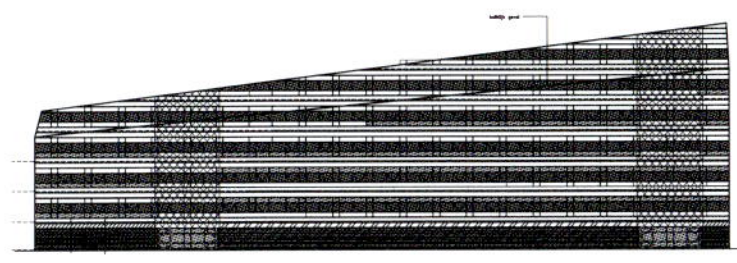

facade 4

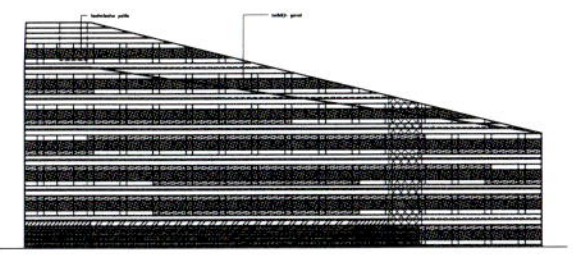

facade 5

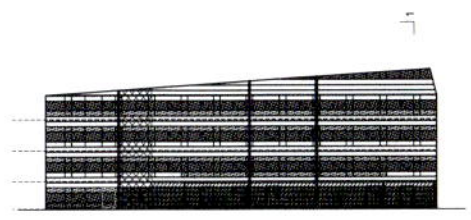

facade 3

\ facade

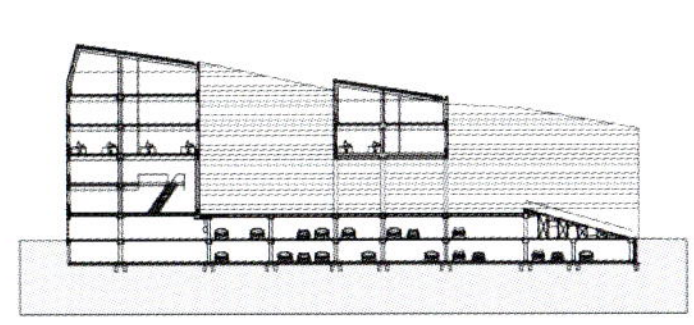

section 2

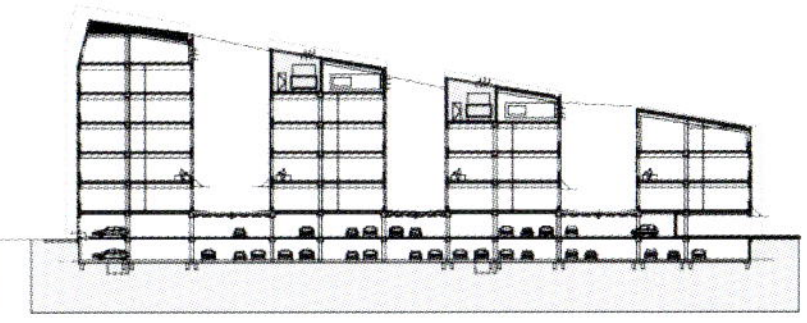

section 3

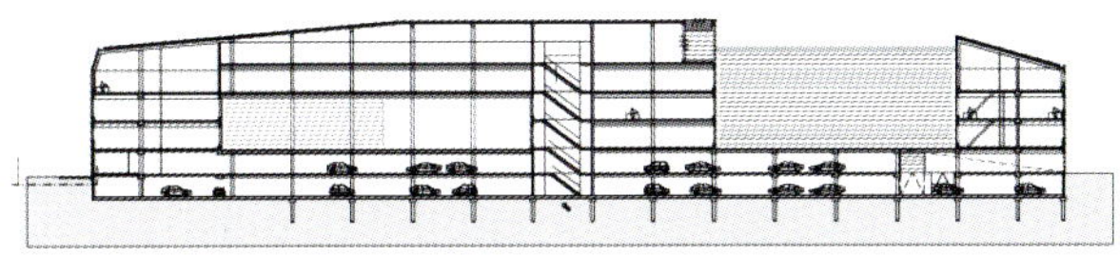

section /

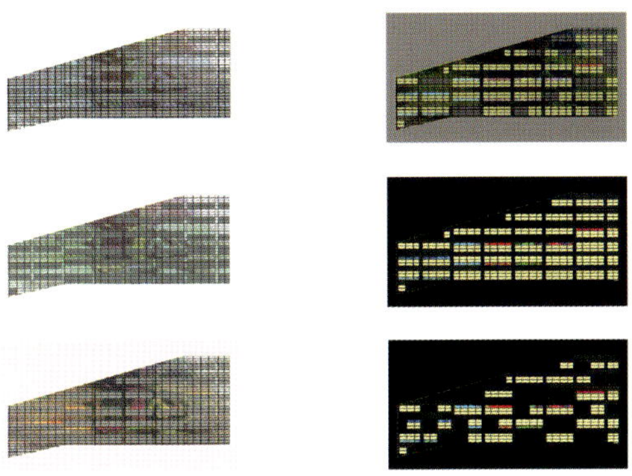

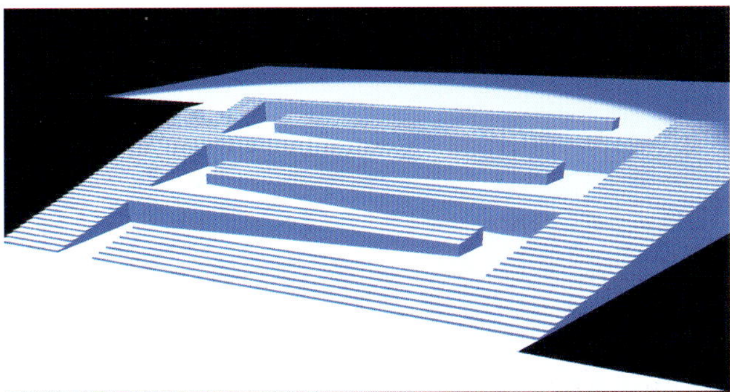

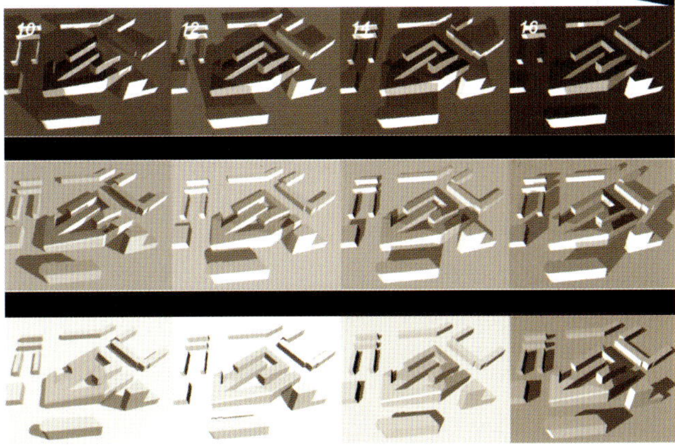

\ zone studies

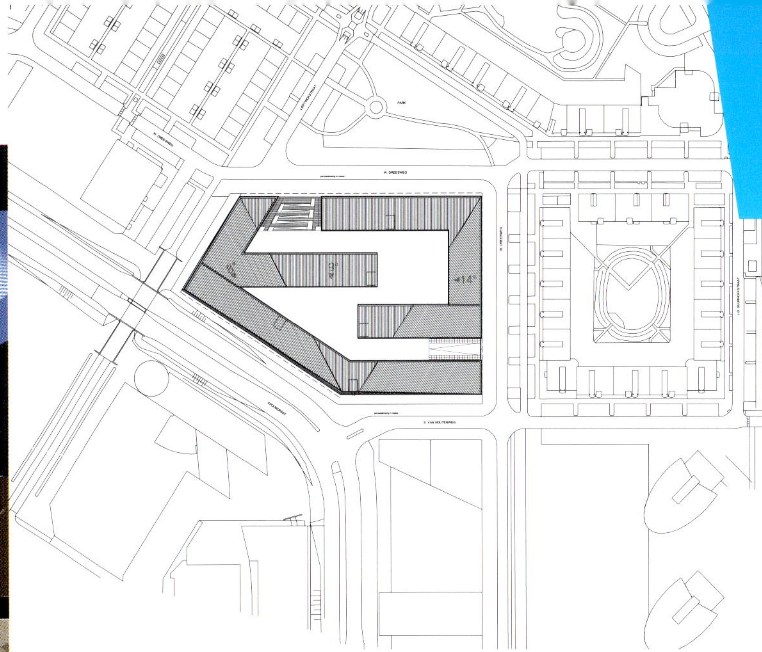

site plan /

3100: 12100 m3 reductie

3500: ca 2500 m3 reductie

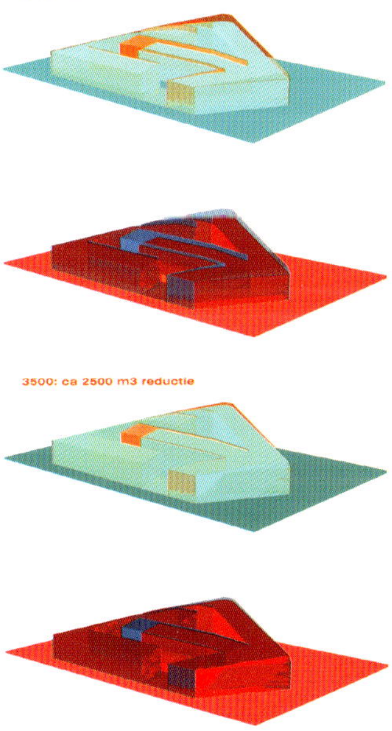

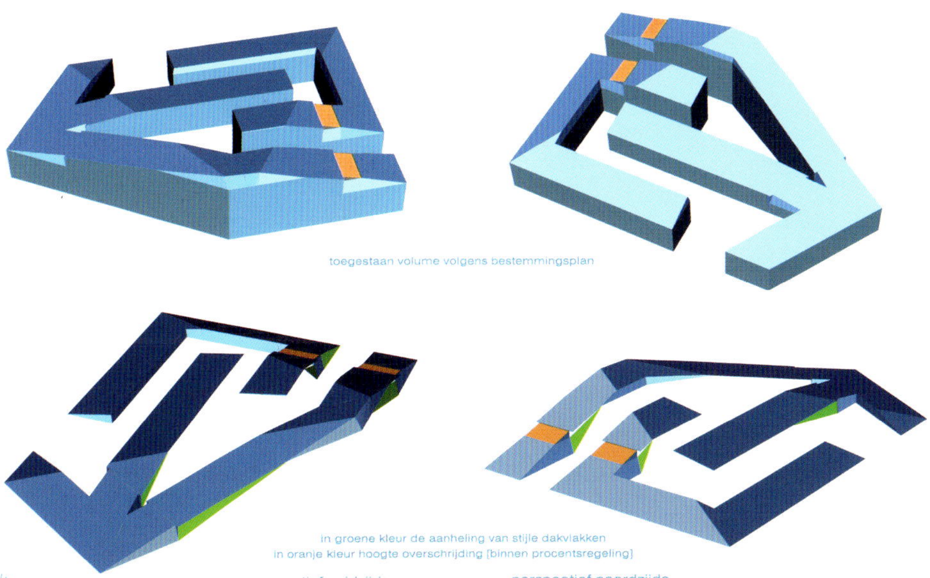

\ volume studies

80.45
S. VAN HOUTENWEG

82.35

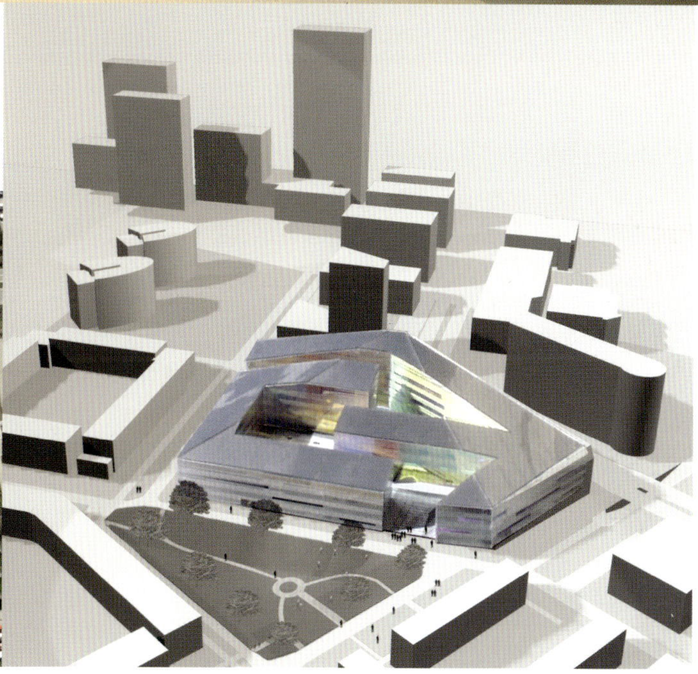

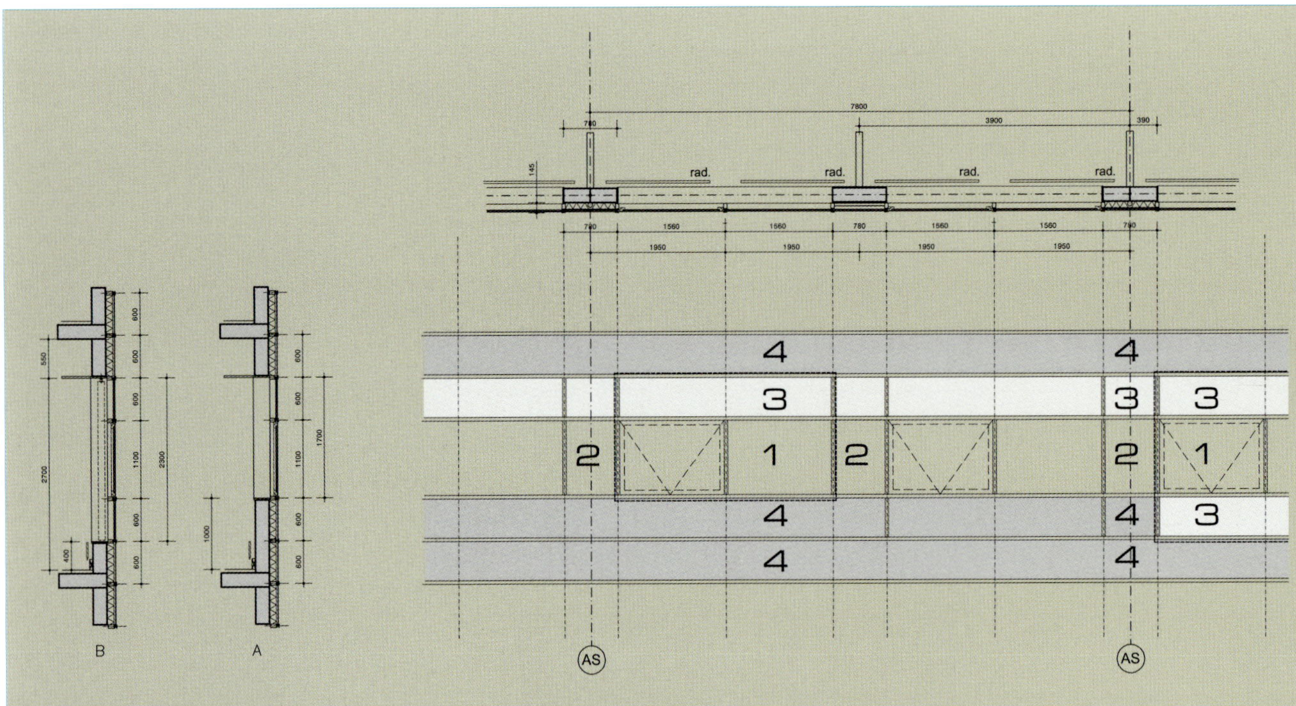

Aluminum (outer) façades consisting of:

1 - See through isolated glass: Glaverbel, Stopray Silver.

2 - Non-see through glass element: similar to Stopray Silver, Glaverbel

3 - Semi transparent: isolated glass: Glaverbel, Stopray Silver, (position 1 seralit screen-print coating)

4 - Non-see through glass element: anodized aluminum plates, mica dual-tone 9706 PVDF, pearl blue, metal rolled pattern

aluminum outer facades

glass inner facades

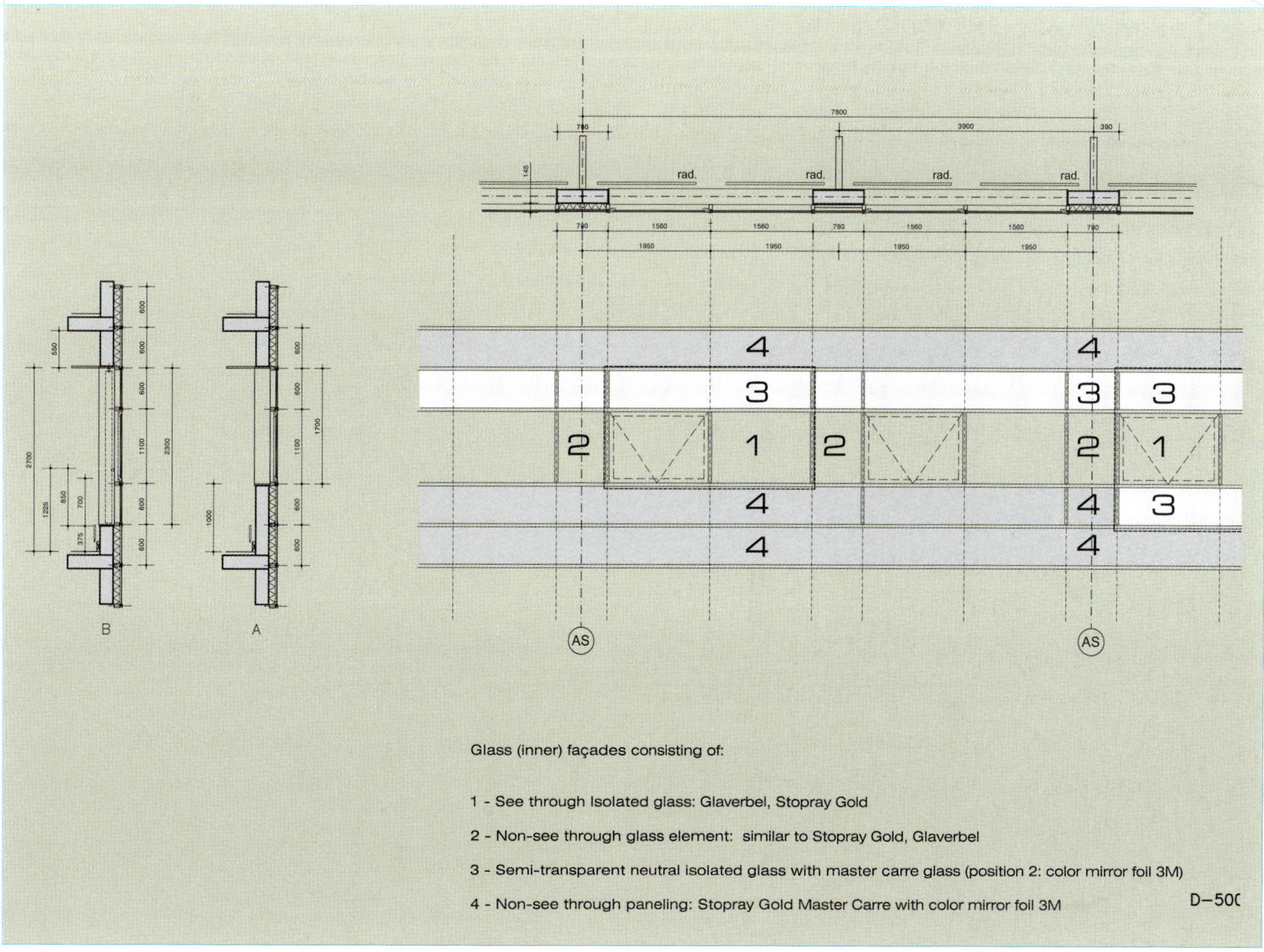

Glass (inner) façades consisting of:

1 - See through Isolated glass: Glaverbel, Stopray Gold

2 - Non-see through glass element: similar to Stopray Gold, Glaverbel

3 - Semi-transparent neutral isolated glass with master carre glass (position 2: color mirror foil 3M)

4 - Non-see through paneling: Stopray Gold Master Carre with color mirror foil 3M

D–50(

Hilversum, the Netherlands

1999-2006

Neutelings Riedijk Architecten

Function_ Offices, Museum, Archives, Audiovisual library

Total floor area_ 30,000m^2

Photographer_ Daria Scagliola

Client_ Netherlands Institute for Sound and Vision

Project architect_ Willem Jan Neutelings, Michiel Riedijk

Structural Engineer_ Aronsohn Raadgevende Ingenieurs

Facade Consultant_ Neutelings Riedijk Architects, Bureau Bouwkunde

Graphic design facade_ Studio Jaap Drupsteen

NETHER-
LANDS
INSTITUTE
FOR
SOUND
AND
VISION

네덜란드 음향 영상 협회는 네덜란드에서 라디오 텔레비전 방송이 시작된 초창기부터 제작된 모든 시청각자료들을 소장하고 있다. 건물은 기록보관소/수장고, 미술관, 사무실, 고객 영접, 서비스 등의 다섯 요소로 구분되어 있다. 이 부분들은 건물의 중심에 있는 '빛 우물'을 둘러싸고 있는데 건물의 공적인 영역을 구성하는 이곳은 리셉션 홀, 복원시설, 영상관 등으로 구성된다. 협회의 모든 시설은 이 거대한 중심공간을 통해 서로 엮이고 있다.

요구된 프로그램의 절반 정도가 엄격한 환경조건에 관한 규정이 적용되지만, 일광이 필요 없는 수장고와 기록보관실이 있다는 점을 고려해 건물을 2개로 수평적으로 분리하기로 했다. 지하부분에는 기록보관실이 있고, 지상에는 미술관과 자연채광이 필요한 다른 용도의 시설들이 있다. 지하공간과 지상공간을 연결해주는 것은 공공 공간인 고객접대영역과 서비스영역이다.

중앙의 '빛 우물'은 지하실의 가장 낮은 레벨까지 일광을 전달한다. 우선 천창을 통해 빛이 유입되고, 다음으로 색을 머금은 완화된 빛이 건물 정면의 유리창을 통해 들어온다. 거대한 '빛 우물'은 남쪽을 향해 개방되어 있는데, 이것은 오후의 빛이 건물의 중심까지 들어오도록 유도하고 반사된 빛이 사무실의 내부 입면 위로 흘러가게 한다.

출입구에서 깊은 협곡처럼 보이는 이 공간을 통해서 방문자들은 수직의 기록보관소/수장고의 엄청난 규모를 편안하게 느끼게 된다. 협곡의 한쪽 면은 평평한 반면, 다른 한 면은 반전된 일련의 테라스로 이루어져 있다. 이 테라스에는 방문객들을 영접하는 방과 기록보관실이나 수장고를 지원하는 별실이 포함되어 있는데, 기록보관실과 수장고는 평평한 협곡 면 뒤에 숨겨져 있는 셈이다.

중앙의 빛 우물은 미술관과 사무실들이 가장 잘 보이는, 거대하게 비워진 이 공간에서 절정에 이른다. 뒤집힌 계단식 우물 같은 미술관의 레벨들은 벽에 설치된 조각 같은 인상을 주는데 이것은 내부공간을 형성하면서 건물 내부 공간을 잘게 나누고 있다.

The Netherlands Institute for Sound and Vision houses all the audiovisual material produced in the Netherlands since the early days of Dutch radio and television.

Our building divides into five distinct elements: archives/stores, museum, offices, client reception and services. Together these parts bound a central well at the building's heart. Here is the building's public portion comprising the main reception hall, restoration facilities and video auditoria. This large central space stitches together all the components of the institute.

Considering that something like half of the required programme encompasses storage and archive rooms with rigorously stipulated climatic conditions but no need for daylight, we decided on a horizontal division into two. The portion below ground contains the archives vault, that above ground the museum and other use forms requiring natural light. Bridging the gap between these two portions are the public spaces, client reception and services.

The central well delivers daylight down to the lowest levels of the vault. In the first instance zenithal light streams in through the skylights; in the second, coloured and tempered light enters through the glazed frontage of the superstructure. The large well in the superstructure opens to the south so that the afternoon sun penetrates to the core of the building and reflected light can skim over the inner facade wall of the offices.

At the entrance the void presences as a deep canyon that dramatically brings home to visitors the scale and the sheer size of the archives/storage vault. One of the canyon's sides is a flush wall, the other rises in a series of inverted terraces. These contain the rooms for receiving clients plus annexes serving the archives and stores; the archives and stores themselves are concealed behind the flush canyon wall.

The central well culminates in an enormous void where both museum and offices show their best face. The upside-down cascade of museum levels registers as a wall sculpture that shapes and scales the internal space of the building.

Neutelings Riedijk Architecten

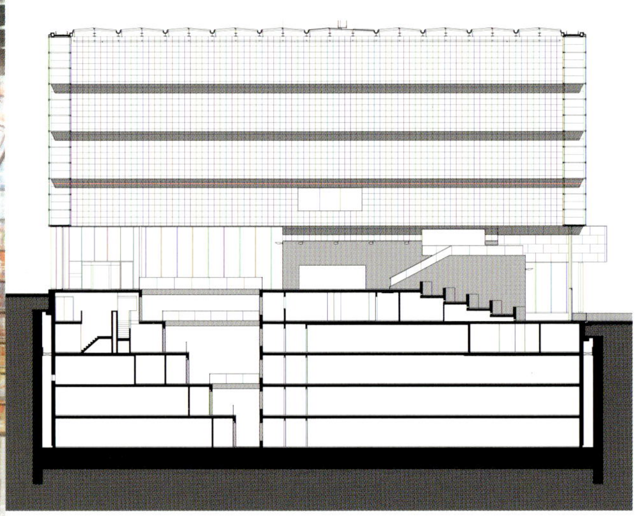

section 1 /

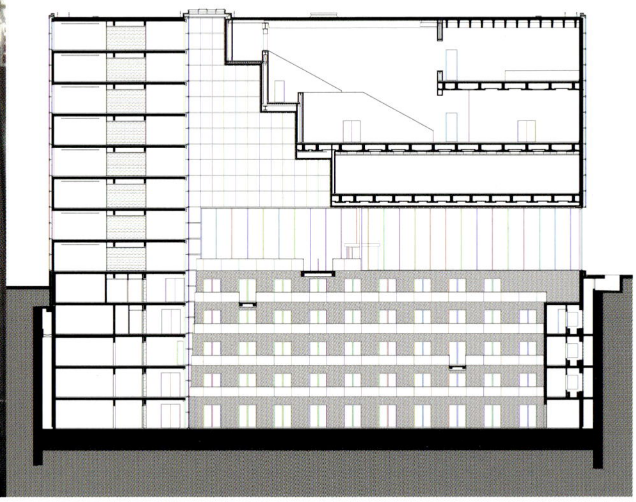

section 2 /

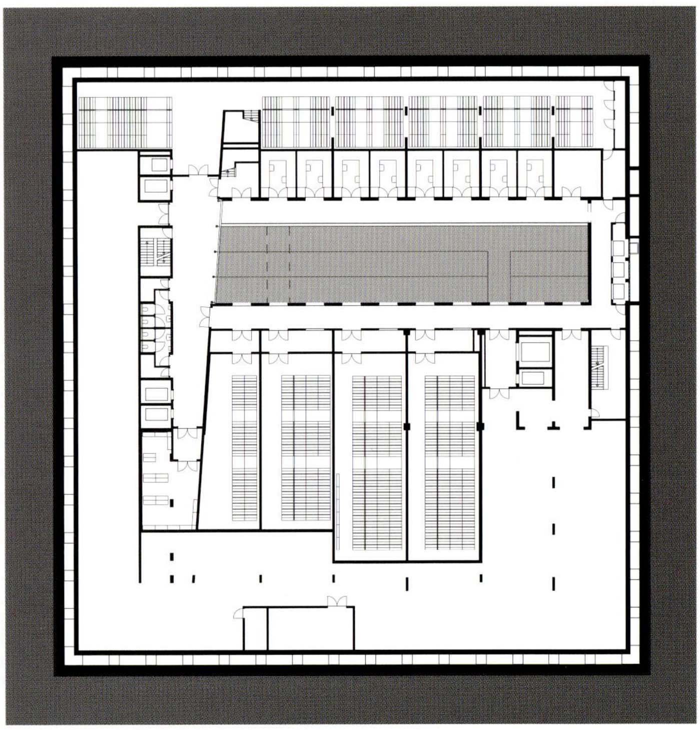

\ layer -2

NETHERLANDS INSTITUTE FOR SOUND AND VISION

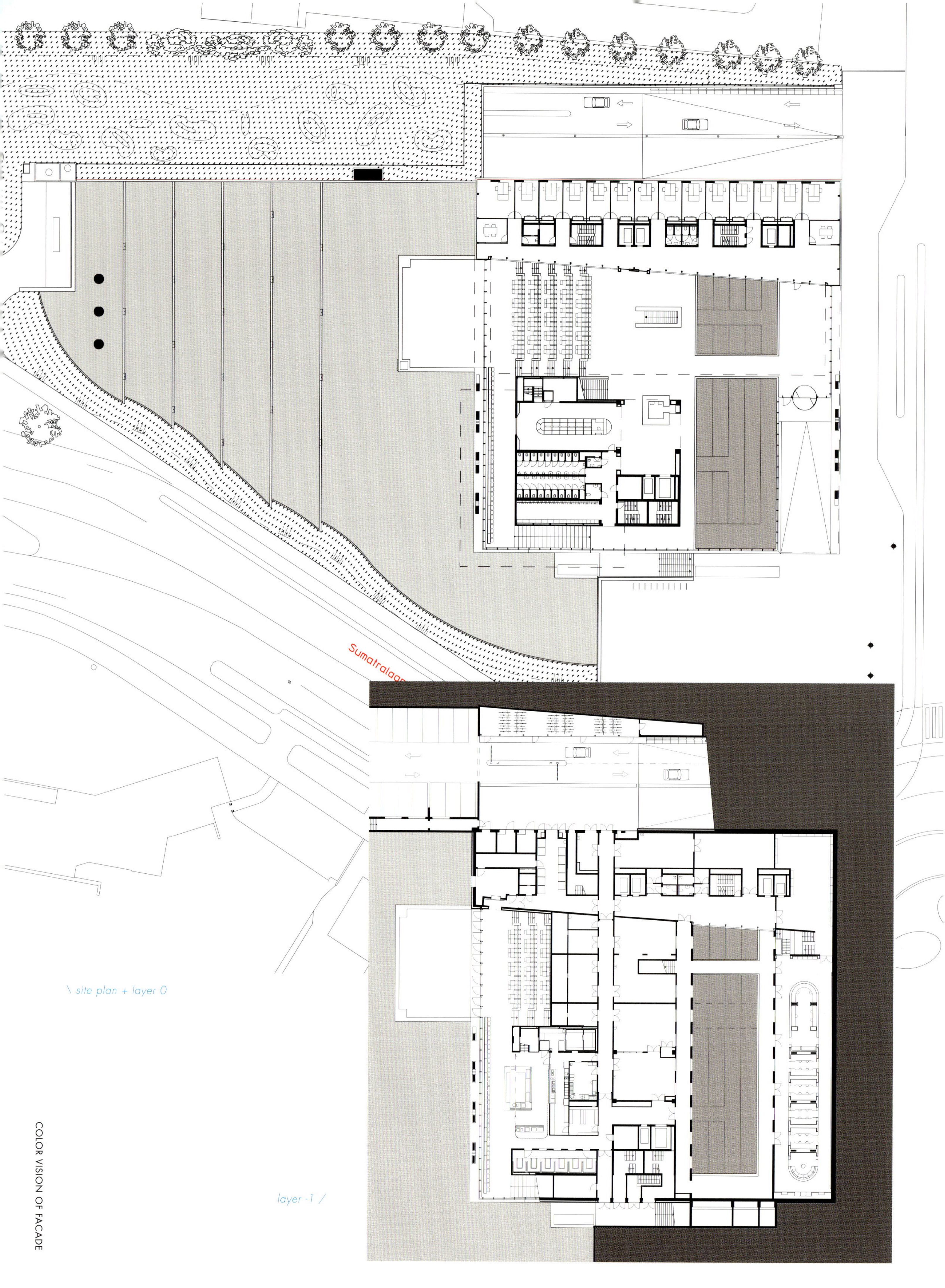

\ site plan + layer 0

layer -1 /

\ layer +5

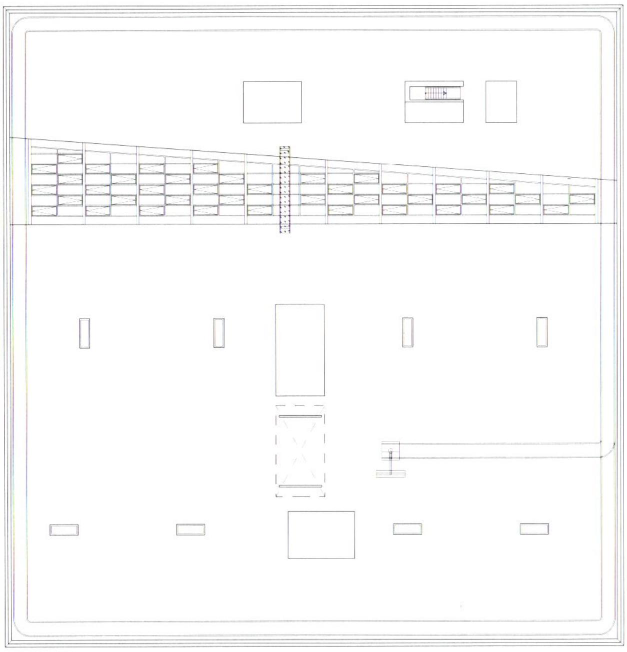

layer roof /

\ layer +2

layer +1 /

유리파사드

네덜란드 힐베르쉼(Hilversum)시에 있는 네덜란드 음향 영상 협회 건물의 유리 파사드를 위해, 우리는 지속가능한 고부조로 만들어진 색 유리 판넬에 네덜란드 tv 역사를 모두 간직하고 있는 협회의 자료들로부터 얻은 tv 이미지 원본을 새길 수 있는 방식을 찾고 있었다. 이러한 방식으로 우리는 성당의 창문에서 빛이 전달되는 것 같은 효과를 얻을 수 있었으며, 반짝이며 비추어지는 기술의 하나이기보다는 쉽게 촉감을 느낄 수 있는 표면을 창조하고자 하였다. 파사드를 만들기 위해서, 우리는 제프 드루스틴(Jaap Drupsteen)과 TNO 아인트호벤(TNO Eindhoven)의 연구팀과 유리 제조사 세인트 고바인(Saint Gobain)과 협력하여, 전체 생산 라인을 발전시키기까지 3년이라는 시간을 보냈다. 그 결과 748가지의 색상이 입혀진 고부조의 이미지를 얻었으며, 이 이미지들은 2100개 이상의 유리패널에 적용되었다. TV로 방영되었던 회사의 업적을 반영하는 광범위하게 다양성을 가진 이미지들을 통합시키고 무작위적으로 선택하여 입면으로 나타내는 것은 일종의 도전이었다. 패널을 제작하는 동안 상당히 많은 선택사항들이 고려되었다. 최종결정에서는 내구성에 비중을 두고, 특별히 이러한 목적으로 고안된 인쇄 방법으로 만들어진 광택이 있는 패널에 세라믹이 적용시키는 스테인드 글라스 기법으로 최종결정이 내려졌다.

드루스틴은 편집된 비디오 영상의 장면들을 패널에 옮겼고, 추가 경비 없이 많고 다양한 이미지들을 옮기는 것이 가능하도록 만들어 주는 완전한 디지털 방식을 적용하였다. 더욱 어려웠던 점은 구부러진 유리에 이용되는 것과 같은 원리에 의존하여 부조방식을 만들어내는 과정이었다. 영상 장면들을 고른 후에, 드루스틴은 그것의 명백한 이미지들 CNC 연마기를 이용해 MDF판에 넣었다. 최초의 CNC는 컴퓨터 수치 제어라는 것을 의미하는 프로그램된 코드를 이용한 컴퓨터 기법의 하나였지만, 이번 경우에는 MDF판의 앞뒤를 가로지르며 끝에 달린 구슬을 조종하는 것을 말했다. 가마에서 그 판을 넣고 원하는 이미지의 나머지 부분을 형틀에 각인시키게 된다. 유리판은 형틀이 놓여 코팅된 채로 820°C에서 가열된다. 이 온도에서 이미지는 태워져 유리 안으로 흡착되며, 유리판은 모양을 만들 수 있을 만큼 무른 상태가 된다. 이렇게 만들어진 색이 입혀진 고부조 방식의 판넬은 자외선이 차단되고, 오랜 시간 내구성을 유지한다.

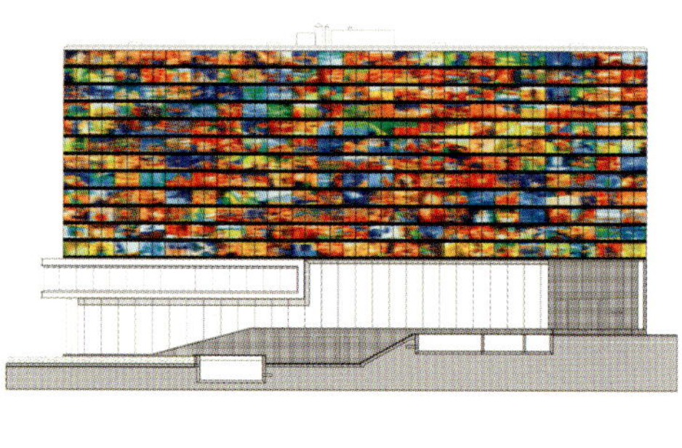

\ east elevation

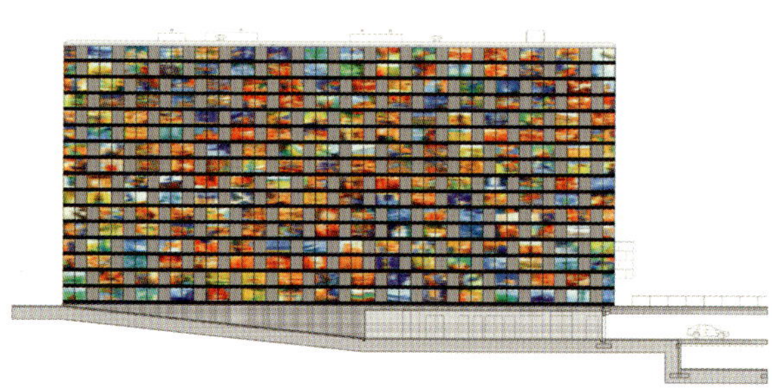

\ west elevation

Glass Facade

For the glass facade of the Netherlands Institute for Sound and Vision in Hilversum, we were looking for a way to translate original tv-images taken from the archives on the Institute which contain the whole Dutch tv history, into sustainable coloured high relief glass panels. In this way we would be able to achieve the quality of light transmitting cathedral windows, which would be more than a piece of glazed and mirrored technology, but instead create a lightly tactile surface.

To make the facade, it took three years for we together with Jaap Drupsteen, in collaboration with a research team from TNO Eindhoven and glass manufacturer Saint Gobain, developed an entirely new production line. The result was 748 coloured, high-relief images, which were applied in more than 2100 glass panels. The challenge lay in incorporating a wide diversity of televised images into the facade, thus presenting a random selection from the institute's archives. A great many options were considered for the manufacturing of the panels. The final decision, based largely on durability, was stained-glass technique in which a ceramic paste is applied to glazed panels by means of a printer developed especially for this purpose.

Drupsteen transferred the manipulated video stills to the panels, a completely digital operation allowed him, theoretically, to transfer an unlimited number of different images at no additional cost. Much more difficult was the process of creating relief, which relies on the same principle used in slumping glass. After choosing a video still, Drupsteen milled its positive images into a MDF panel with a CNC milling machine. The initials CNC stands for computer numerical control, a computer technique with a programmed code that, in this case, guided a ball-tipped bit back and forth across the MDF panel. He then placed the wood panel, coated on one side in ceramic paste, was laid on the sand mould and the oven heated 820°C. At this temperature, the paste burned the image into the glass, and the glass panel softened enough to take on the shape of the mould. Thus creating a coloured, high relief glass panel; that is UV resistant and incorporates a long term durability.

Neutelings Riedijk Architecten

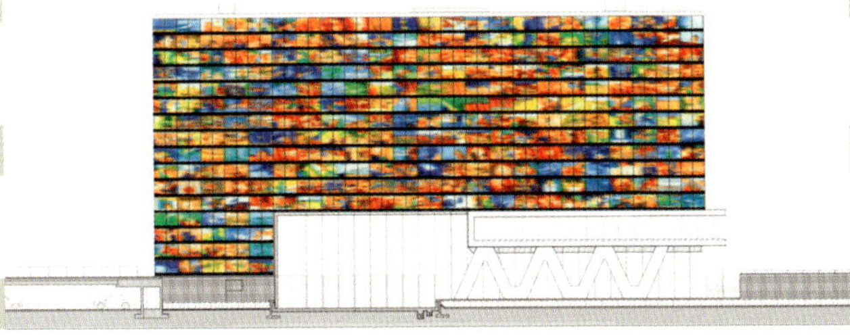

\ south elevation

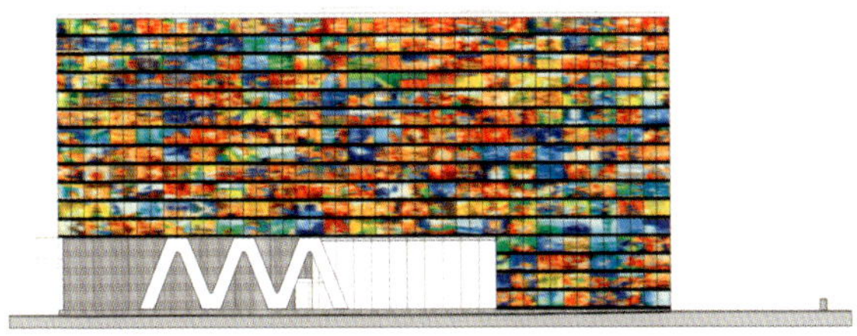

\ north elevation

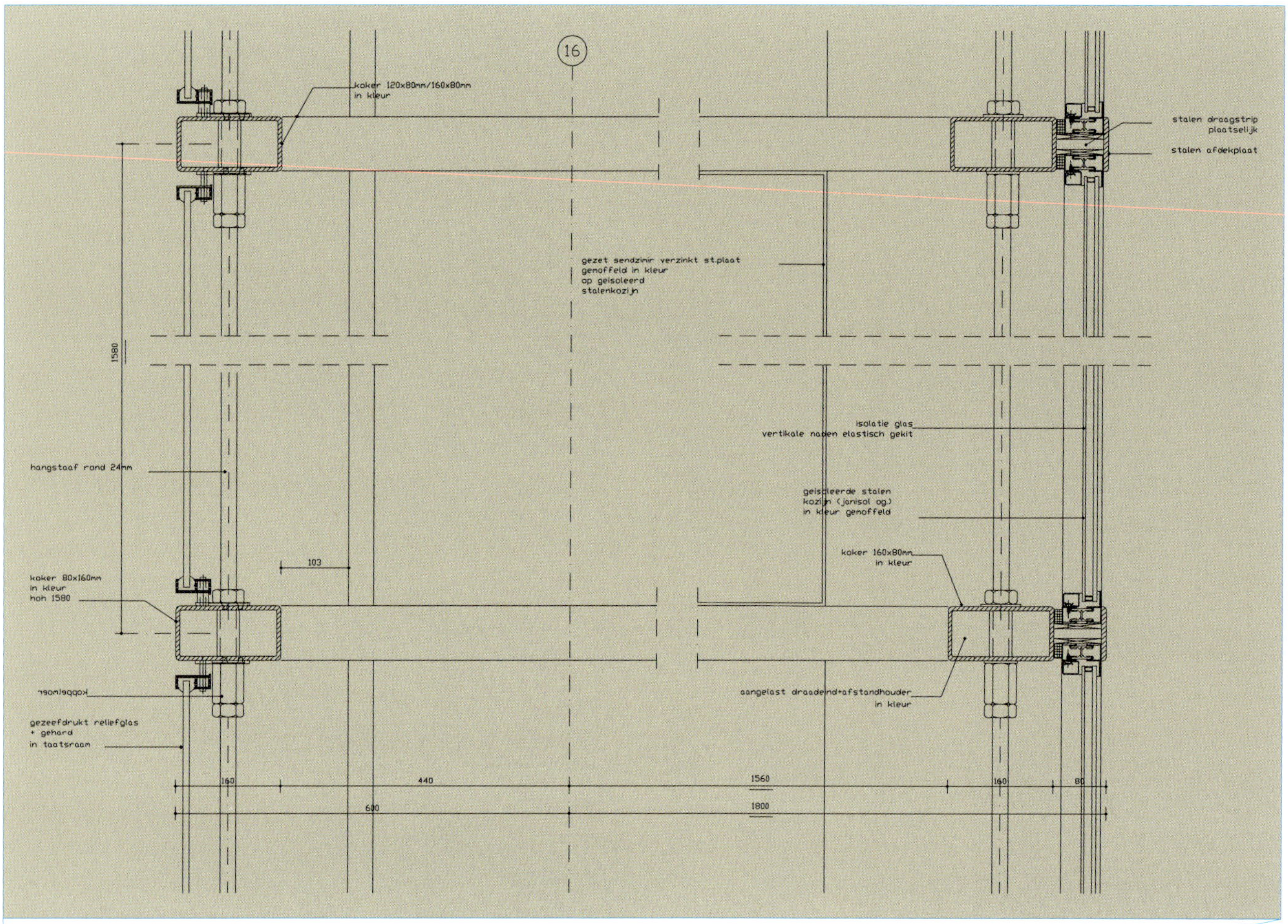

south facade detail

office facade detail

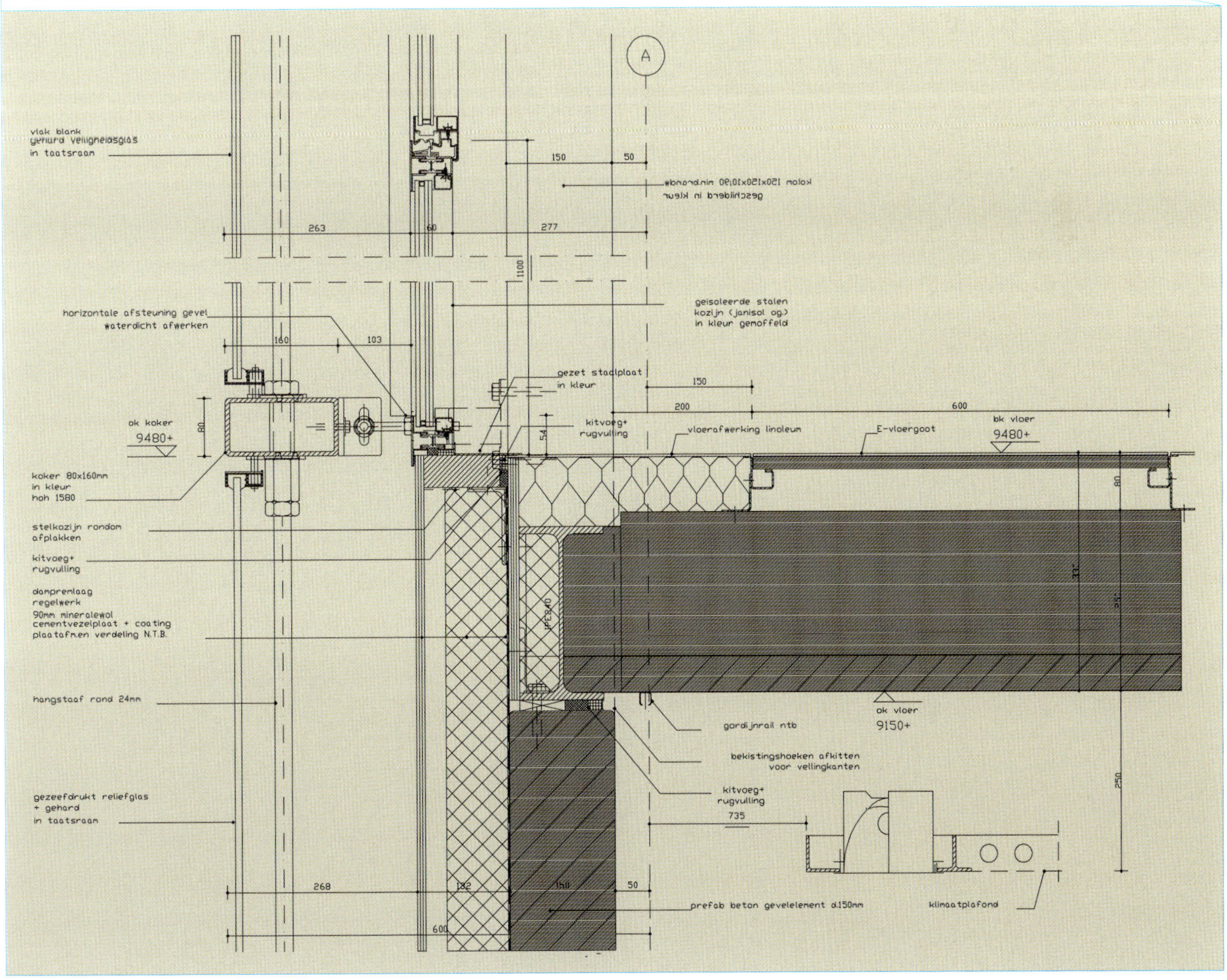

Doetinchem, the Netherlands

2006

EEA

Function_ School

Total floor area_ 16,400m², learning(8,922m²), offices(529m²), gym(1,955m²), canteen(300m²)

Photographer_ Christian Richters

Client_ Stichting CoVoa, Doetinchem

Project architect_ Erick van Egeraat

Structural Engineer_ ABT-C, Delft

Mechanical Engineer_ ABT-I, Velp

Facade Consultant_ Alkondor, Hengelo

METZO COLLEGE

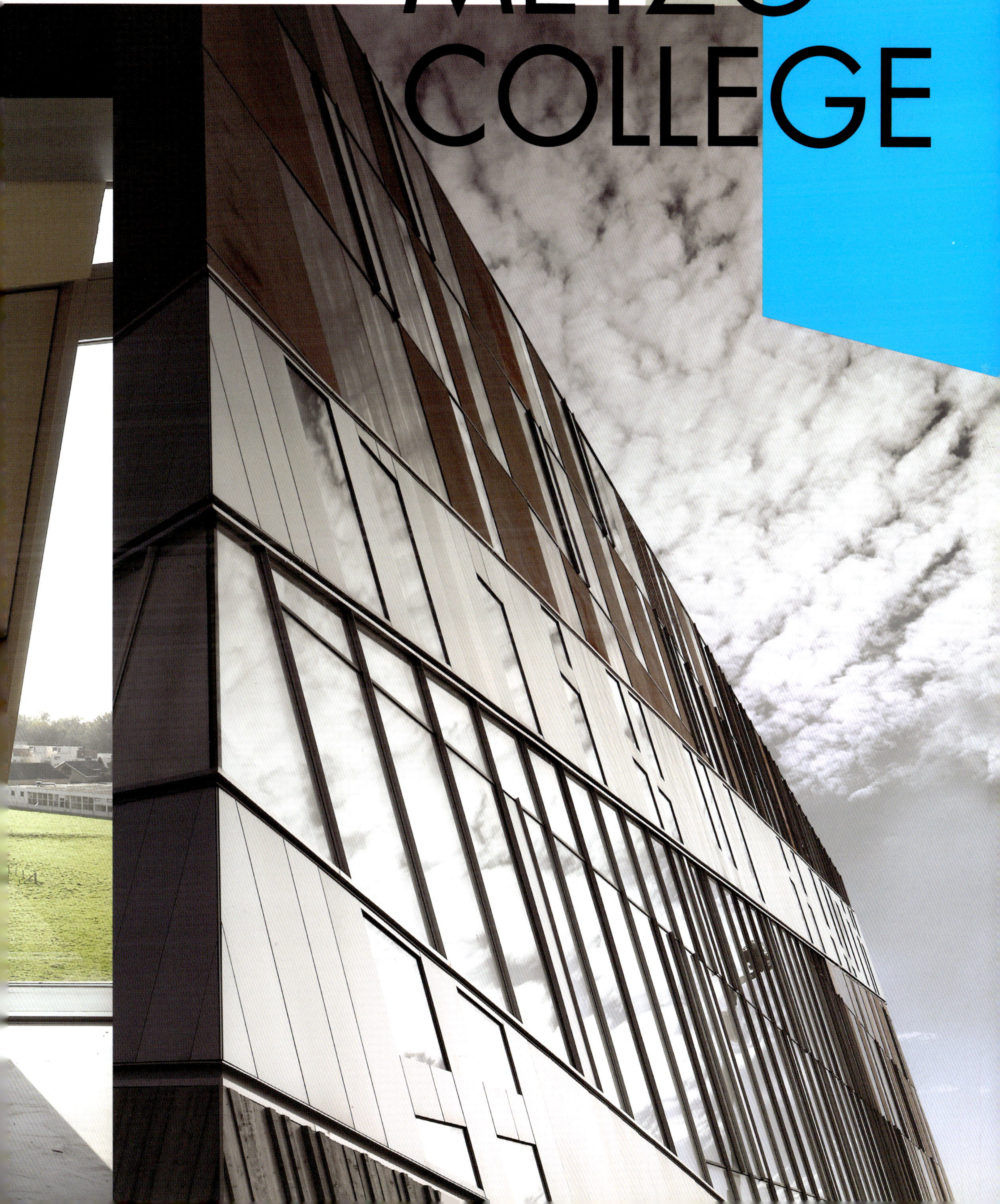

네덜란드 서부 도시 도팅헴(Doetinchem)에 위치한 멧조 대학은(Metzo College)는 공공 스포츠 시설을 비롯한 직업교육과정을 수용하고 있다. 새로운 건물은 기존 학교의 건물 3개동을 대체한 것이다. 학교는 총 연면적이 16.000m²이며, 12세에서 16세 사이의 학생 1,300명을 수용하고 있다. 또한, 보건, 복지 교육과정을 포함한 기술과 이론 과정들을 제공하고 있다.

이번 프로젝트는 교육 공간을 만들고, 직업교육시스템 내의 여러 학생 집단 간에 더 나은 사회적 유대를 형성하는 것을 도울 수 있는 건물을 만드는 것을 목적으로 출발하였다.

일반인들이 이용할 수 있는 운동 시설은 학교의 사회적 특성을 증대시키면서 여느 시설과 다른 공공성을 지닌 교육 시설이 되도록 한다.

건물은 개방된 녹지 위에 놓여있다. 설계상 주요한 목적은 탁 트인 주변 환경을 최대한 보전하는 것이었다. 그 결과 주변 경관에 통합되면서 동시에 주변 환경을 존중하는 집약된 피라미드 형태의 건물이 되었다. 여기에 더하여, 집약된 건물은 내부의 평면계획과 방향성도 최적화하여 교육과정의 사회적 측면이 향상되도록 한다. 이러한 건물의 개념은 6층 규모의 집약된 볼륨 안에 포함되어 있다.

주출입구가 3층에 놓이도록 지형을 들어 올렸다. 주출입구는 3층 높이에 놓이는데, 이 중앙부분을 통해 학생들의 건물 내부 순환동선이 최적화된다. 방문자들은 이 출입구로부터 식당이나 체육관 같은 다른 공공시설로 접근할 수 있다. 출입구 레벨 상부에 놓인 대형 파티오가 바닥을 절개한다. 지붕에 매달린 파티오의 설치를 통해 일광이 건물 내부로 유입되도록 했다. 덕분에, 건물 내부에서 길을 찾기도 분명해졌다.

'블랙박스'로 설계된 체육관은 들린 지형 하부의 낮은 레벨에 출입구와 함께 위치한다. 기준층에서는 파티오의 주변으로 개방형 열람실, 개인 열람코너, 그리고 교사들을 위한 워크스테이션들이 놓여있다.

교실들은 건물의 파사드 주변에 배치되었다. 덕분에 실제 생활 조건하에서의 융통성 있는 레이아웃을 실험해 볼 수 있었다. 목표로 삼은 것은 학생들과 교사들에게 구별할 수 있는 작업공간을 제공하고 최대한의 융통성을 가진 공간을 효율적으로 사용할 수 있도록 하는 것이었다. 그 결과 주변 환경 속에 강력한 인상을 주는 집약된 건물이 되었다.

이 건축물은 한 편으로는 사람들을 초대하고, 다른 한 편으로는 안정감을 제공한다.

The new Metzo College, located in Doetinchem, a City in the East of the Netherlands, houses a school for vocational education as well as public sport facilities. The new building replaces three existing school buildings. On a total gross floor area of 16.000m² the school accommodates 1.300 students between ages 12 to 16. The school offers a range of technical and theoretical courses, including health and welfare education.

The project has been initiated in order to realize a building which offers space for education and provides for better social cohesion between the different student groups that exist within the vocational education system. The integration of the publicly accessible sports facilities increases the social character of the school and makes it an educational building with an unusually public character.

The building is located in an open green area. The main design objective was to preserve the existing open surroundings as much as possible. This resulted in a compact pyramid shaped building, which integrates into, and respects the surrounding landscape. In addition, the compact building mass also optimizes the internal planning and orientation within the building and it enhances the social aspect of this type of education.

The concept comprises a compact volume of 6 stories. The landscape has been raised to allow for the main entrance to be located on the third level. This central location optimizes the students circulation within the building. From this entrance visitors have access to other public functions such as the restaurant and the gym. A large patio, placed above the entrance level, cuts the floor. The placement of the patio, which is hanging from the roof, allows daylight to enter into the building. It furthermore allows for clear orientation within the building.

The gym, which is designed as a 'black box', is located on the lower levels with the entrance beneath the raised landscape. On the typical floor plans, open study areas, individual study corners and workstations for teachers are located along the patio. Classrooms are located along the facade of the building. This results in a flexible layout in which real life situations can be simulated. The goal was to offer identifiable workspaces for students and teachers and to allow for efficient use of space with a maximum of flexibility. The result is a compact building with a strong identity within the landscape.

The building is inviting on the one hand and provides a sense of security on the other.

Erick van Egeraat

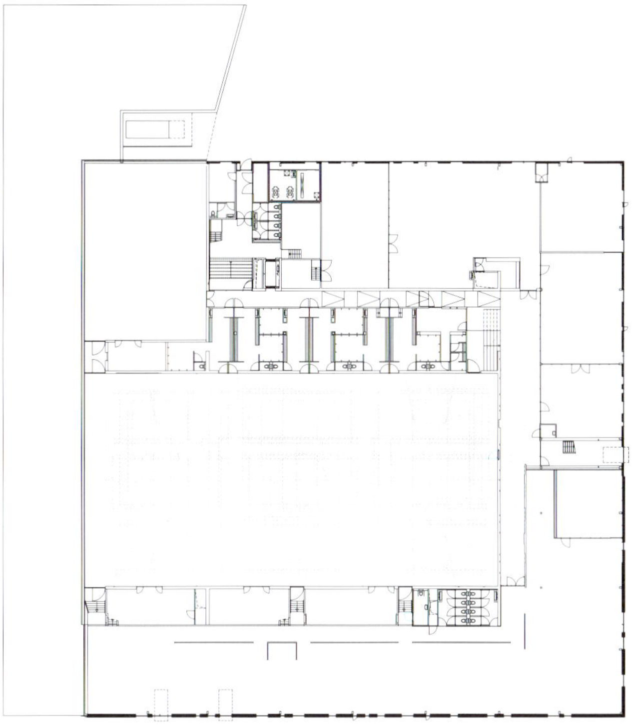

\ ground level

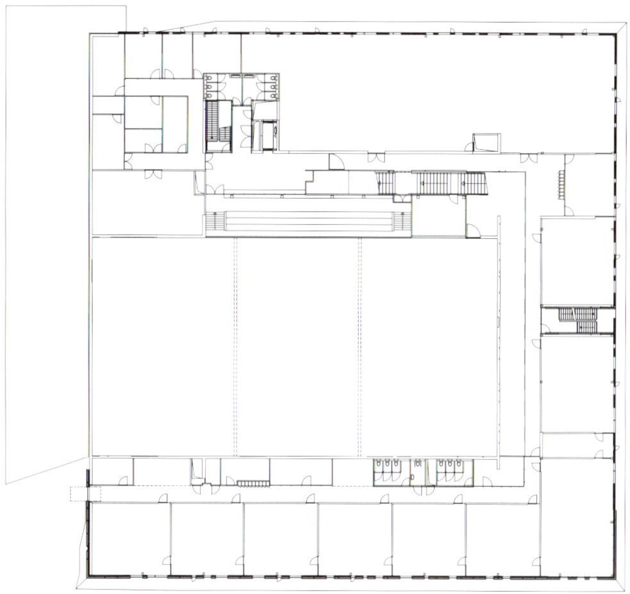

1st level /

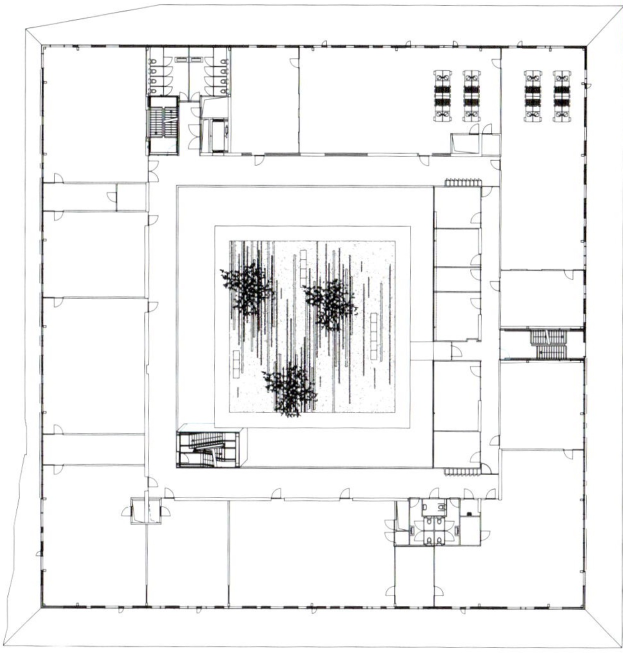

\ 3rd level

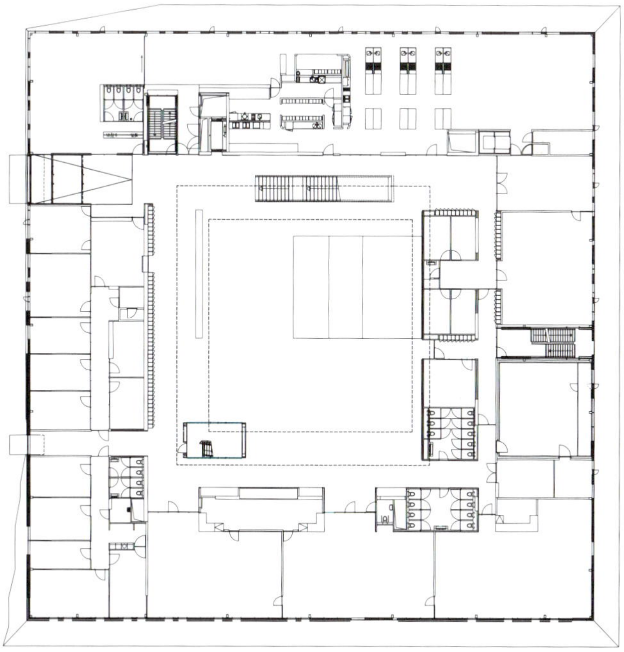

\ 2nd level

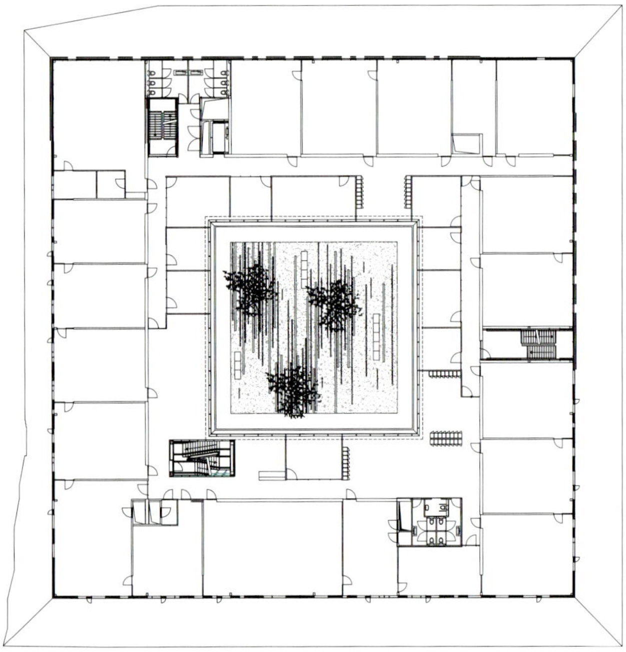

\ 4th level

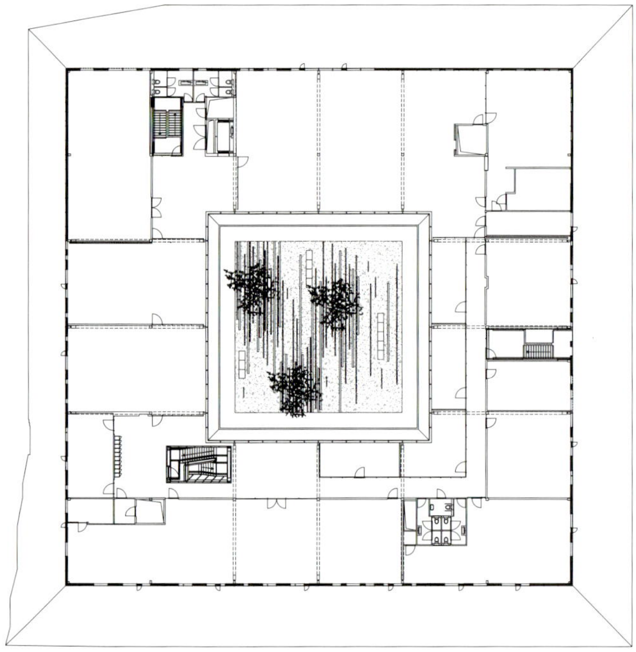

5th level /

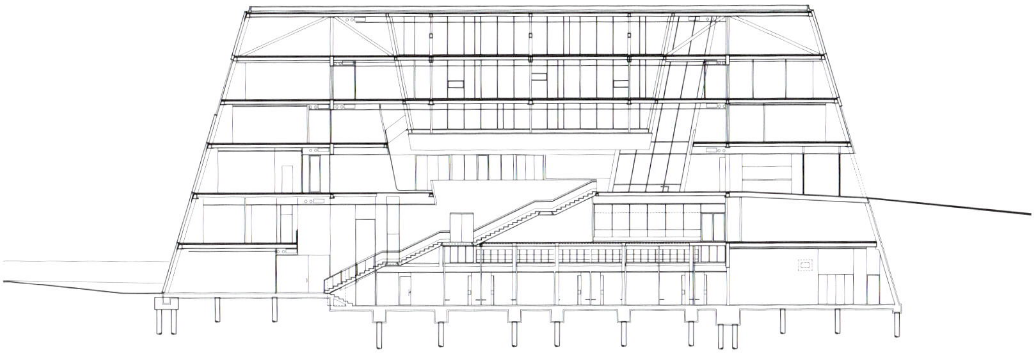

\ section AA

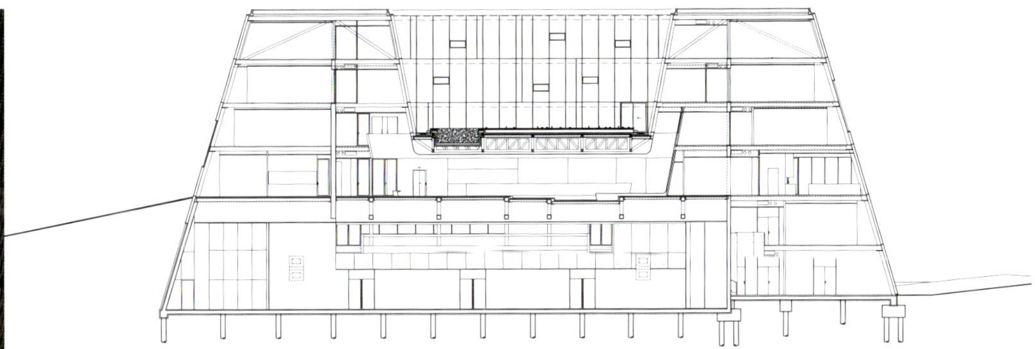

\ section BB

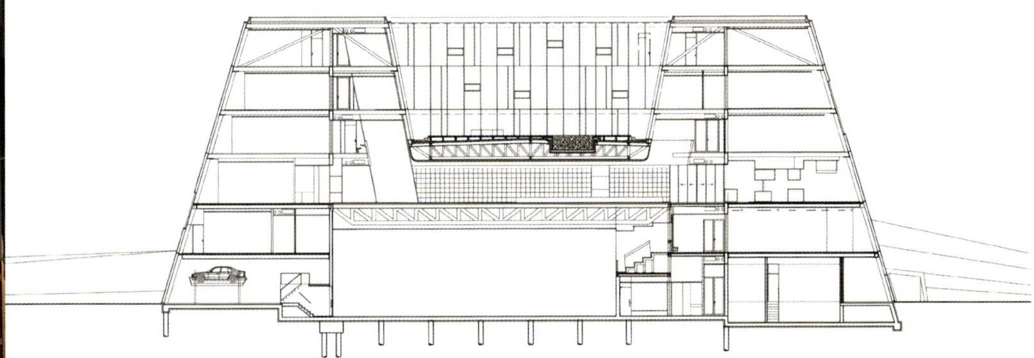

\ section CC

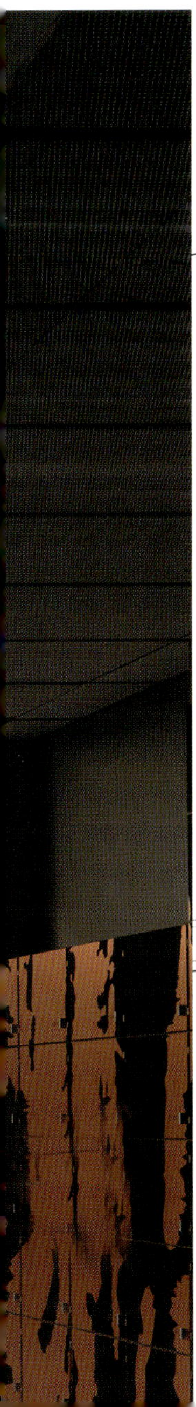

COLOR VISION OF FACADE

Since the end of World War II the Free University has occupied a central role in the intellectual life of Berlin. It is one of the city's most symbolically important institutions, its foundation marking the rebirth of liberal education there after the war. Today, with more than 60,000 students, it is the largest of Berlin's three universities. This redevelopment scheme includes the restoration of its Modernist buildings and the design of a new library on the campus.

The University was founded in 1949 in Dahlem, West Berlin, by the Western Allied Powers to compensate for the loss of the Humboldt University historically Germany's premier seat of learning - which was isolated in the Soviet Sector following the division of Berlin. The University's web-like campus was designed in 1963 by the architects Candilis, Josic, Woods, Schiedhelm and the first buildings were completed in 1973.

The facade, designed by Jean Prouve, followed Le Corbusier's Modulor proportional system and consisted of framed panels in Corten steel, which has self-protecting corrosive qualities. Its rusty appearance led to the affectionate nickname of 'die Rostlaube(the rust bucket)'. In the thin sections used here, however, the Corten steel was prone to decay, which by the late 1990s had become extensive.

As part of a comprehensive process of renovation the steel panels and framing have been replaced with patinated bronze elements. While these are faithful to the original architects' intentions some details have been sensitively altered to meet contemporary technical requirements and energy standards. The roof will be covered with vegetation to add insulation and improve microclimatic conditions.

Six of the University's courtyards have been united to form the site of a new library for the Faculty of Philology. The five-storey building is housed within a free-form skin consisting of aluminium panels, ventilation elements and double-layered glass panels, supported on steel frames with radial geometry. The library's curved form and its double skin create pressure differentials in the cavity between the skins, which assist a natural ventilation system. An inner membrane of glass fibre allows soft sunlight to penetrate the space while creating an atmosphere of concentration. Scattered transparent openings punctuate this membrane at strategic points to allow glimpses of the courtyard.

Foster + Partners

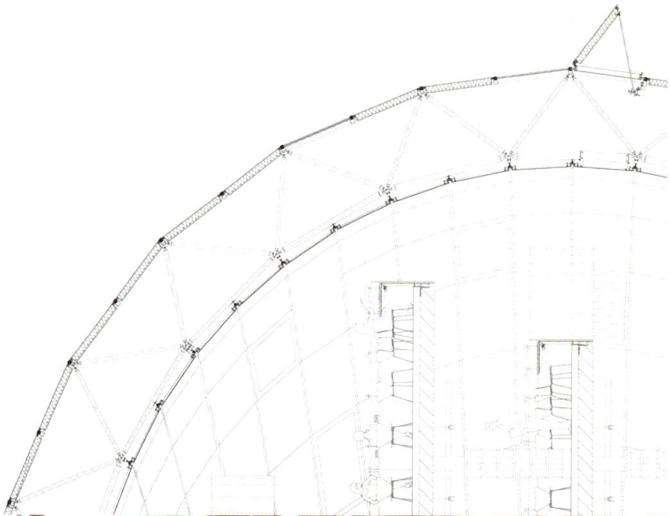

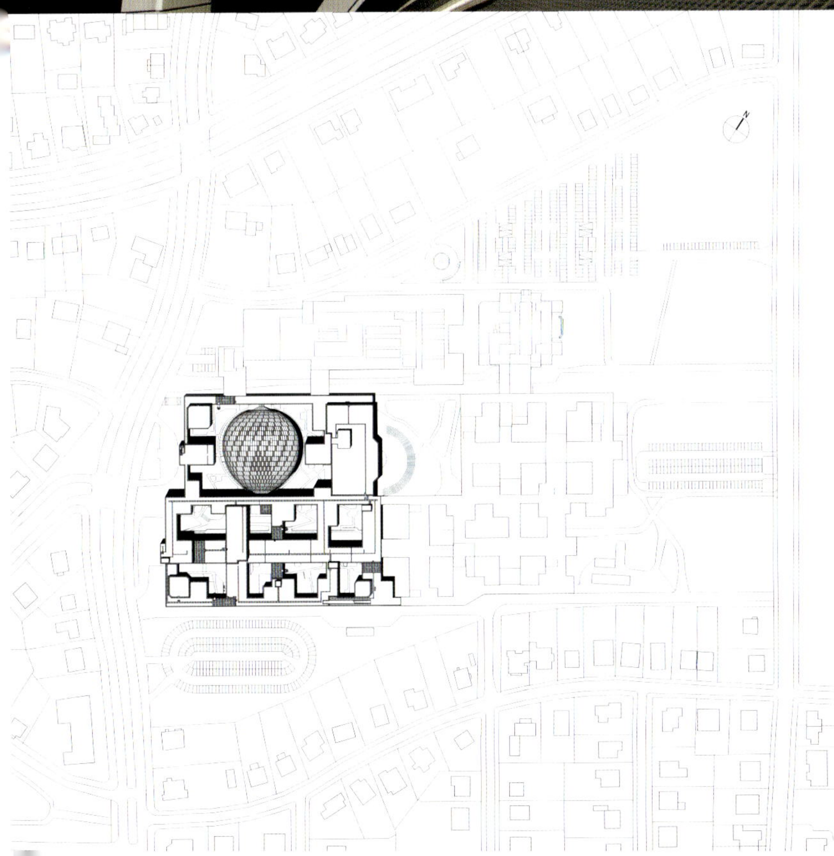

\ site plan

\ sketch

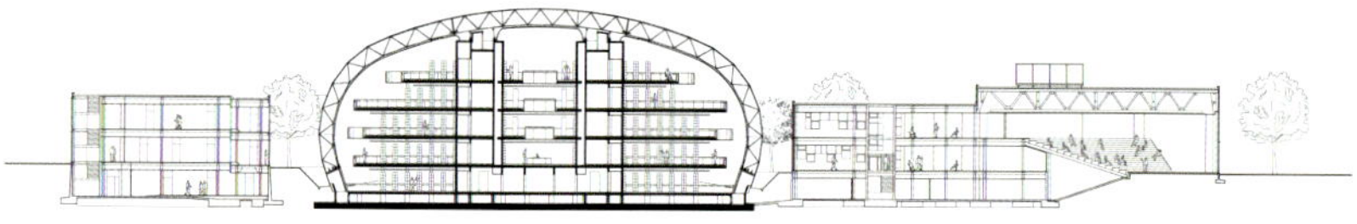

section 1 /

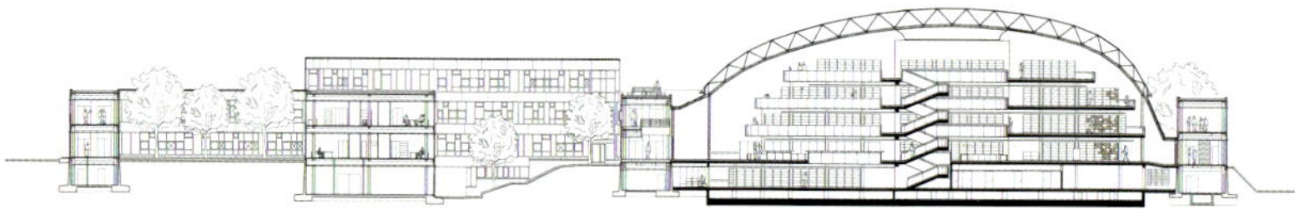

section 2 /

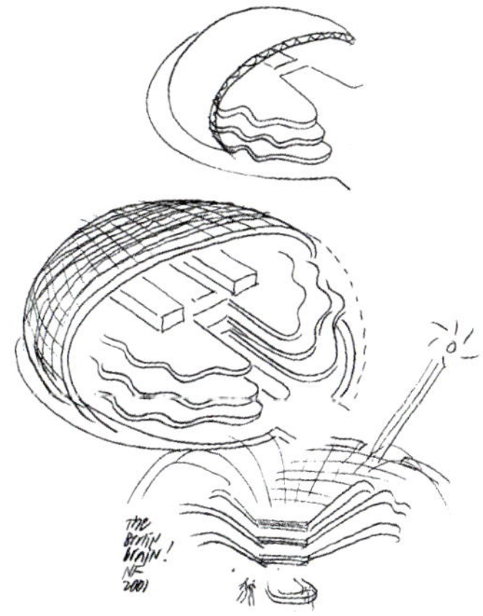

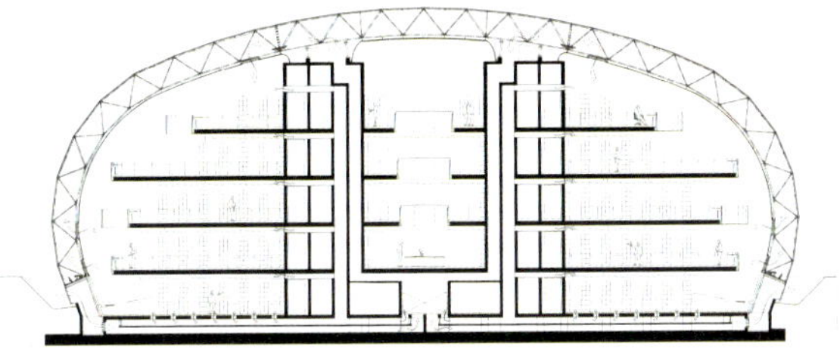

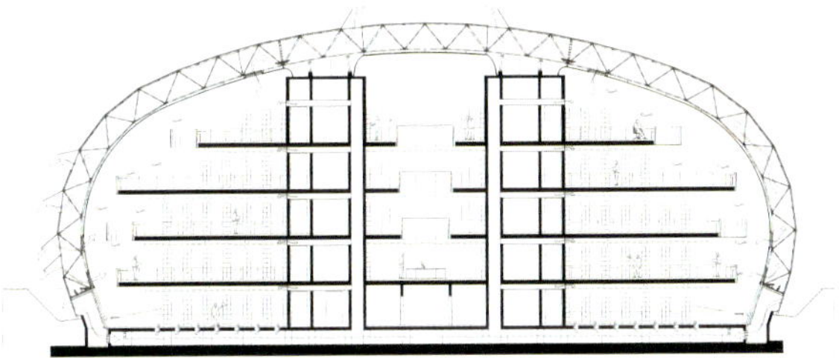

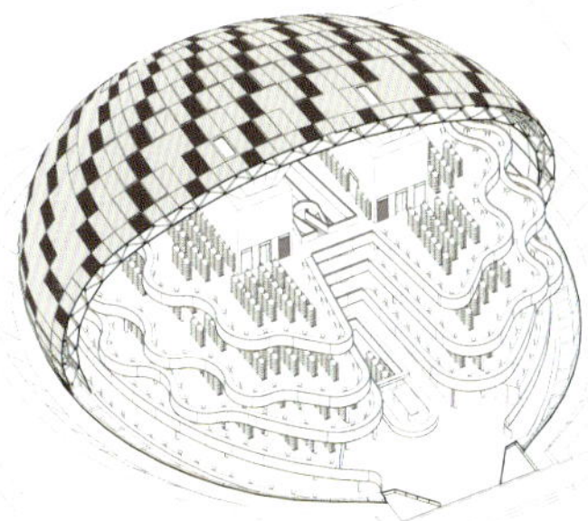

isometric /

FREE UNIVERSITY

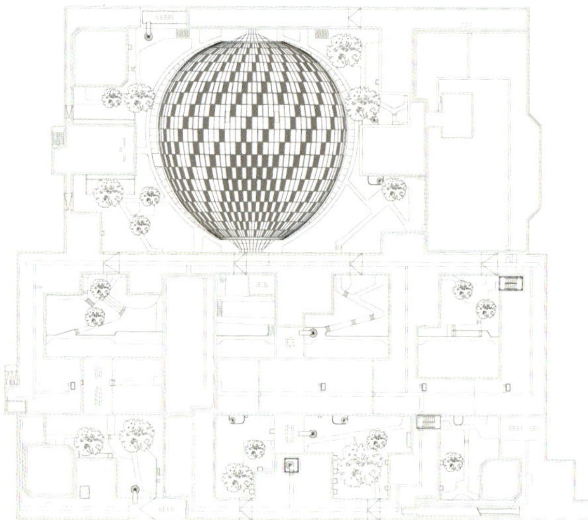

level 6 /

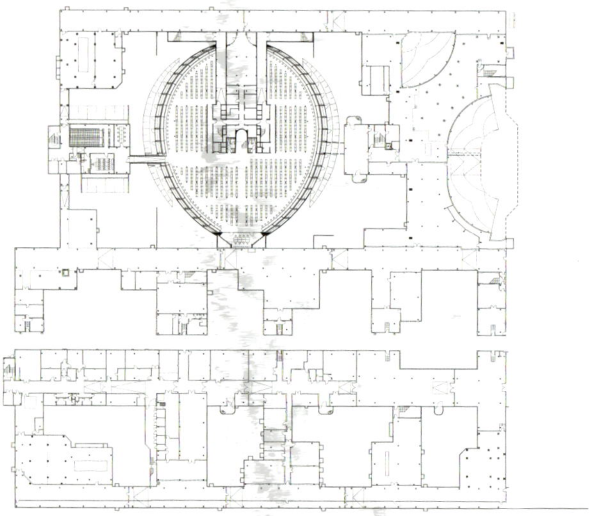

\ level 0

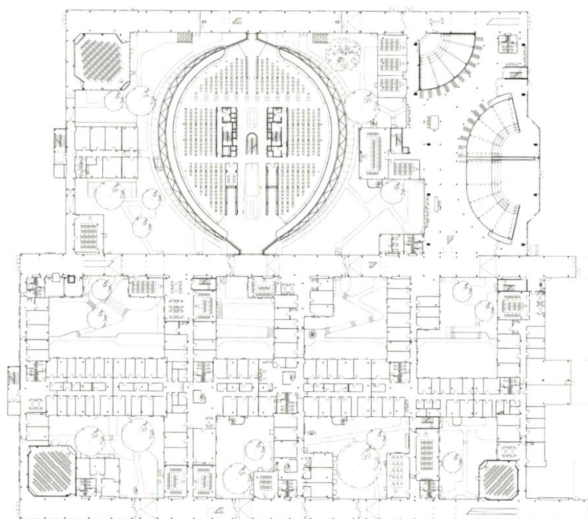

level 1 /

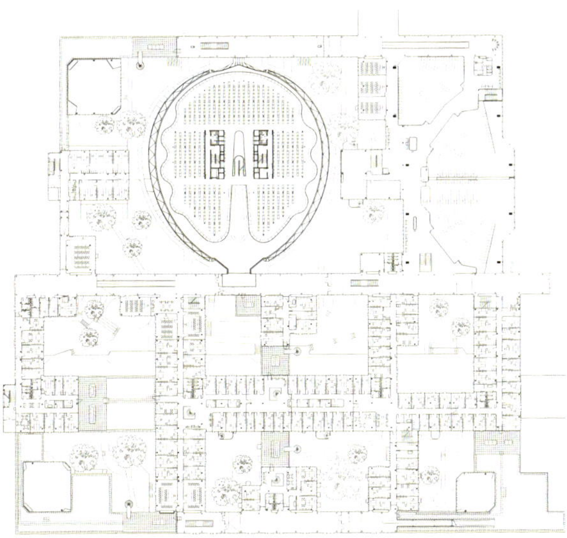

\ level 2

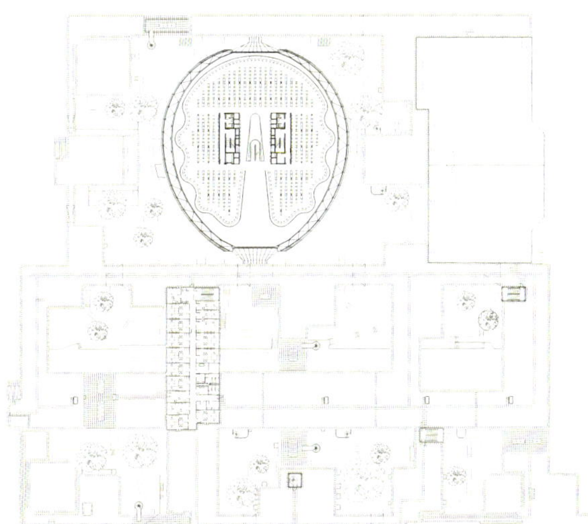

level 3 /

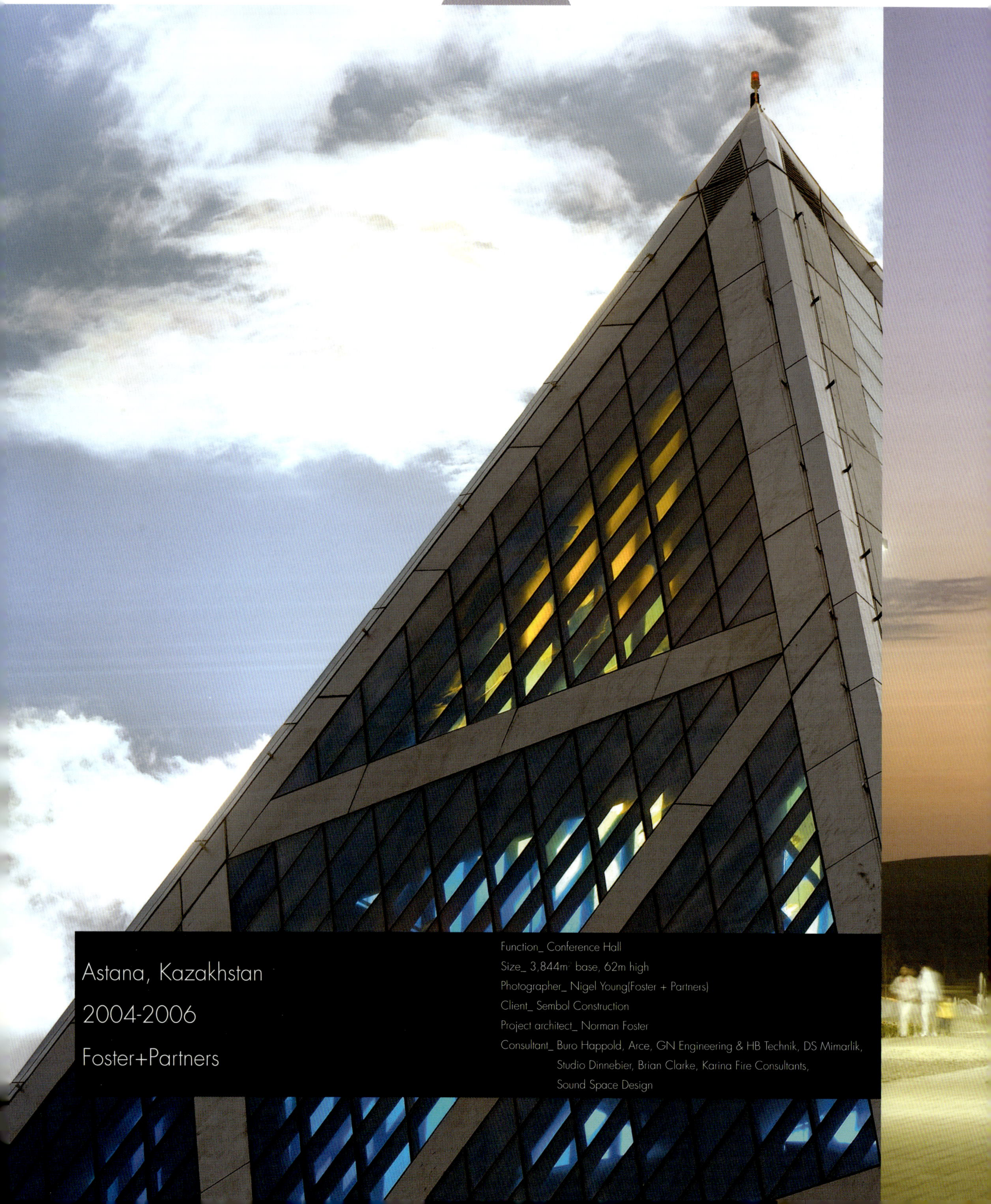

Astana, Kazakhstan

2004-2006

Foster+Partners

Function_ Conference Hall

Size_ 3,844m^2 base, 62m high

Photographer_ Nigel Young(Foster + Partners)

Client_ Sembol Construction

Project architect_ Norman Foster

Consultant_ Buro Happold, Arce, GN Engineering & HB Technik, DS Mimarlik, Studio Dinnebier, Brian Clarke, Karina Fire Consultants, Sound Space Design

PALACE OF PEACE

지난 2003년 9월, 옛 소련 연방 중 가장 큰 나라인 카자흐스탄(Kazakhstan)의 새로운 수도인 아스타나(Astana)에서 세계 지도자들과 종교인들의 회의가 열렸다. 이번 행사의 성공에 고무된 카자흐스탄의 대통령은 이 행사를 3년마다 개최하기로 했다.
국제현상설계경기에 따라, 노먼 포스터 사는 이러한 회의를 위한 영구적인 장소인 평화의 전당을 설계하게 되었다. 이 건물은 종교적 상호 이해의 중심지로서, 폭력을 종식시키고 믿음과 평등을 촉진하는 범세계적인 중심지가 되고자 한다.
지구 상의 모든 종교적 신앙을 대표함은 물론, 평화의 전당에는 1,500석의 오페라 하우스, 문명을 연구하는 대학시설, 그리고 카자흐스탄의 다양한 인종과 지역 단체를 위한 국립 센터가 자리 잡고 있다. 프로그램의 다양성은 62미터의 높이와 62미터 X 62미터의 바닥으로 이루어진 순수한 피라미드 안에서 통합된다. 석재 마감으로 내부의 다양한 기능을 암시하는 창문이 나열된 피라미드의 꼭대기에는 브라이언 클라크(Brian Clarke)의 스테인드글라스 작품이 있다.
이 작품은 공간적으로 상승하는 중앙 아트리움을 둘러싸고 있는데, 화려한 색조의 조명 패턴이 생기를 불어넣고 있다. 회의실은 '평화의 손'이라고 불리는 네 개의 기울어진 기둥에 의지하여 최상층에 올려졌다. 내부의 경사벽을 따라 상승하는 승강기들은 조경과 함께 어우러져 '아스타나의 공중정원'이라 불리는 영접공간으로 대표자들을 인도한다. 여기에서 참석자들은 나선형의 경사로를 이용해 회합실로 올라가게 된다. 아트리움 바닥에 고정된 렌즈모양의 거대한 유리창이 오페라하우스의 객석으로 빛을 드리우고, 건물의 가장 낮은 곳에서부터 최상층까지 수직적 연속성을 만들어내고 있다.
아스타나의 기후는 어려운 해결과제인데, 연교차가 심해서 여름에는 40°C, 겨울에는 영하 -40°C까지 차이가 벌어진다. 공사일정은 대단히 촉박해서, 2006년에 열리는 두 번째 회의에 맞추어 전당의 공사가 완료되어야 했다. 이런 조건 때문에 설계팀은 겨울 동안 현장에서 떨어진 곳에서 제작되고 여름 동안 설치될 수 있는, 미리 제조된 자재를 활용한 구조적 해결 방안을 개발하였다. 프로젝트 브리핑에서 완공까지 모든 과정이 단, 21개월 만에 이루어졌다.

In September 2003, Kazakhstan, the largest of the former Soviet Republics, hosted the inaugural Congress of Leaders of World and Traditional Religions in the new capital, Astana. Spurred by its success, the President of Kazakhstan decided to make it a triennial event. Following an international competition, the practice was commissioned to design a permanent venue for the Congress - the Palace of Peace. The building is conceived as a global centre for religious understanding, the renunciation of violence and the promotion of faith and human equality.

In addition to representing all the world's religious faiths, the Palace houses a 1,500-seat opera house, a university of civilisation, and a national centre for Kazakhstan's various ethnic and geographical groups. This programmatic diversity is unified within the pure form of a pyramid, 62 metres high with a 62 x 62-metre base. Clad in stone, with glazed inserts that allude to the various internal functions, the pyramid has an apex of stained glass by the artist Brian Clarke. Spatially, it is organised around a soaring central atrium, which is animated with spectacular cast light patterns. The assembly chamber is elevated at the top, supported on four inclined pillars, 'the hands of peace'. Lifts rise up the inward leaning walls to take delegates to a reception space lined with vegetation, 'the hanging gardens of Astana', from where they ascend to the chamber via a winding ramp. A broad glass lens set in the floor of the atrium casts light down into the auditorium of the opera house and creates a sense of vertical continuity from the lowest level of the building to the very top.

The Astanian climate posed a significant challenge, with an annual range from 40°C in summer to -40°C in winter. The construction schedule is extraordinarily rapid, the Palace is to be completed in time for the Meeting of the Second Congress in 2006. This led the design team to develop a structural solution that utilises prefabricated components, which can be manufactured off site during the winter months and erected during the summer. The entire process, from briefing to completion will have taken just twenty-one months.

Foster + Partners

site plan /

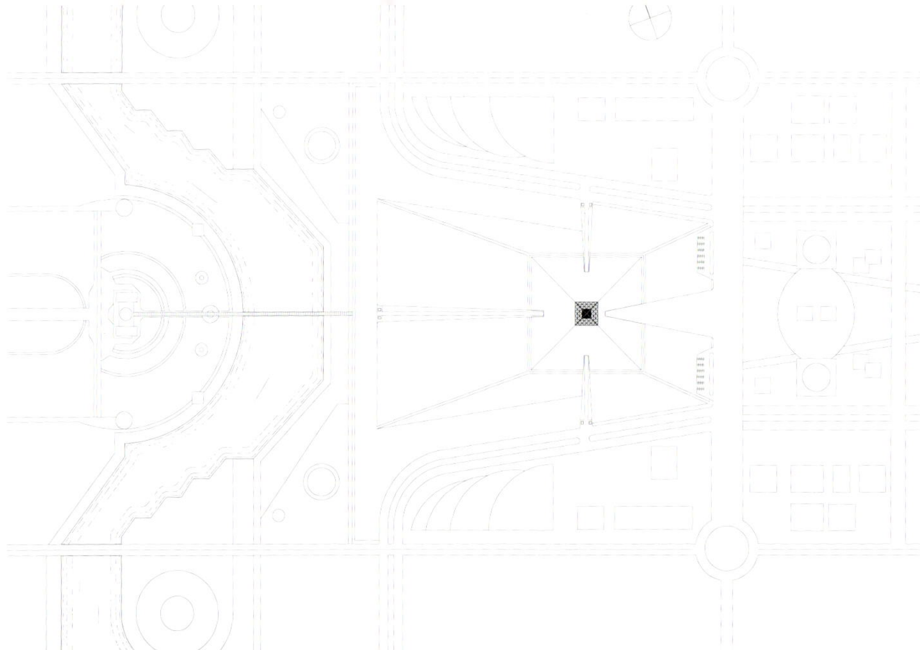

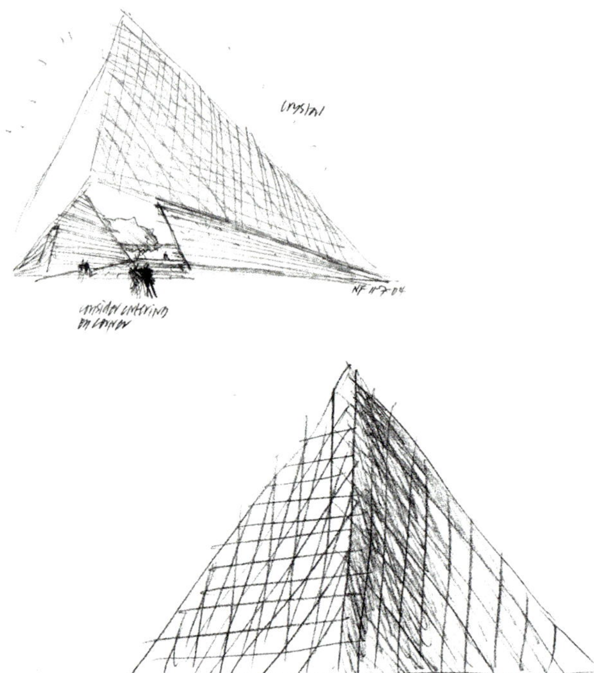

sketch /

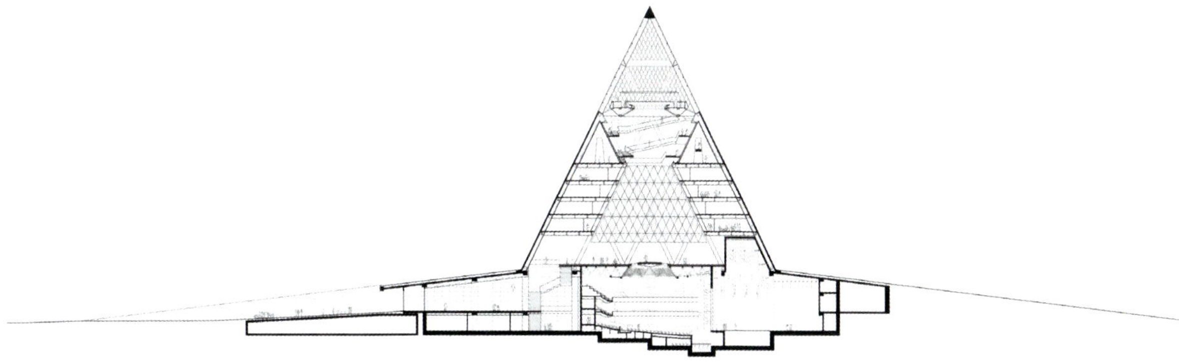

\ section 1

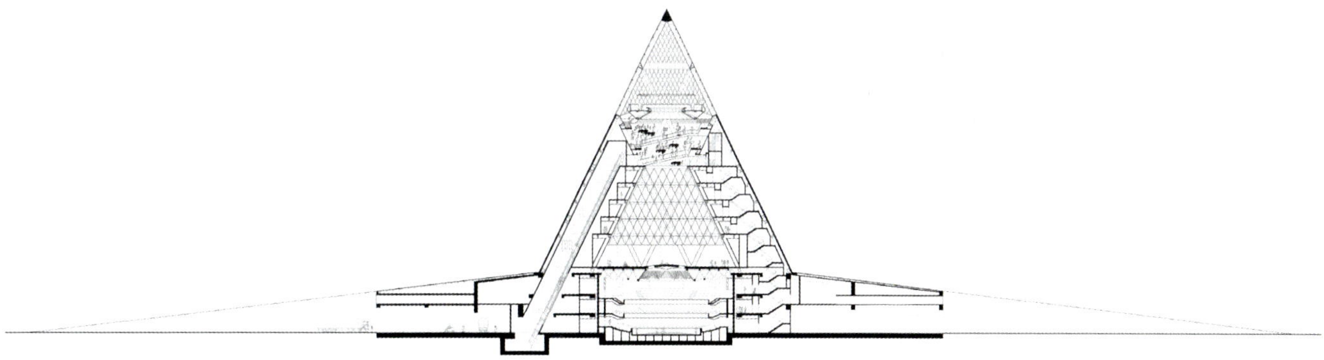

\ section 2

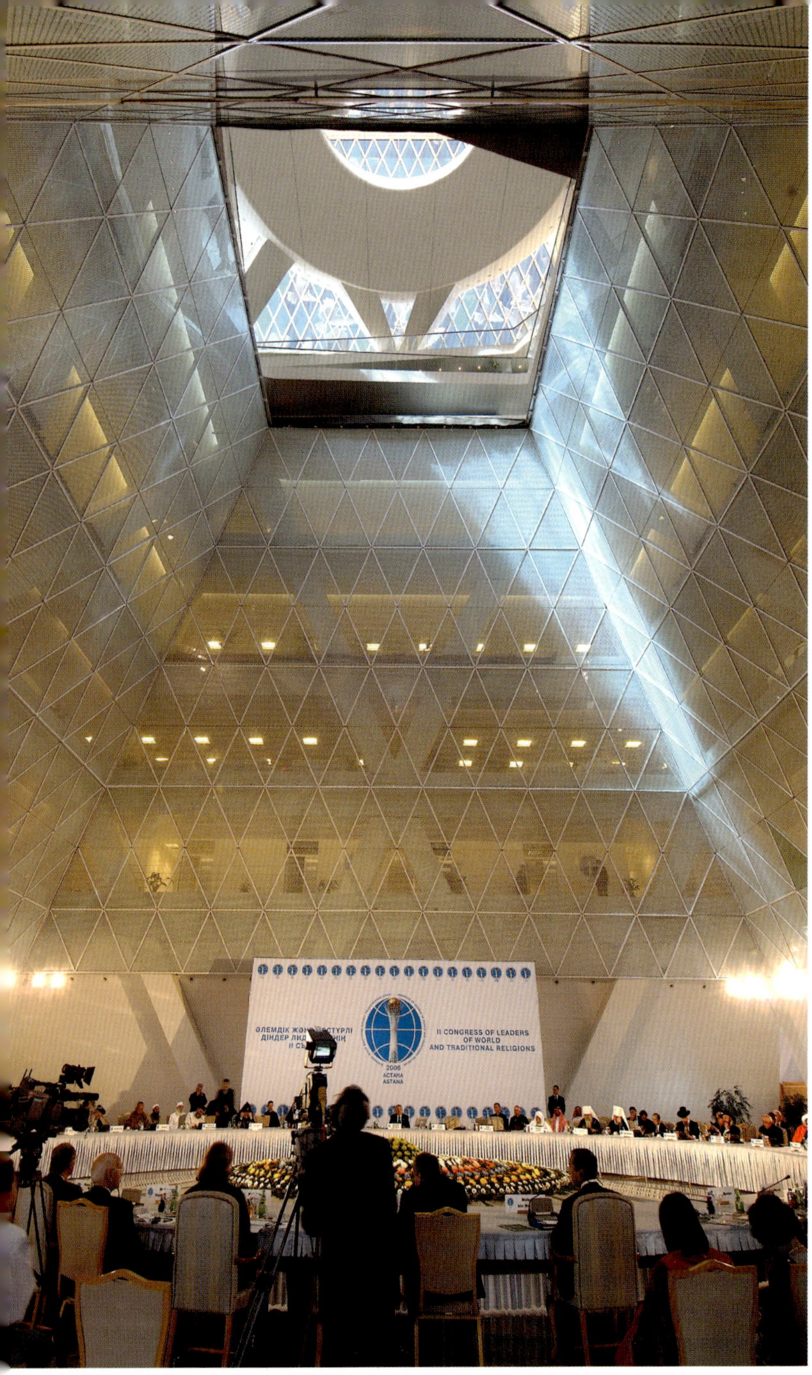

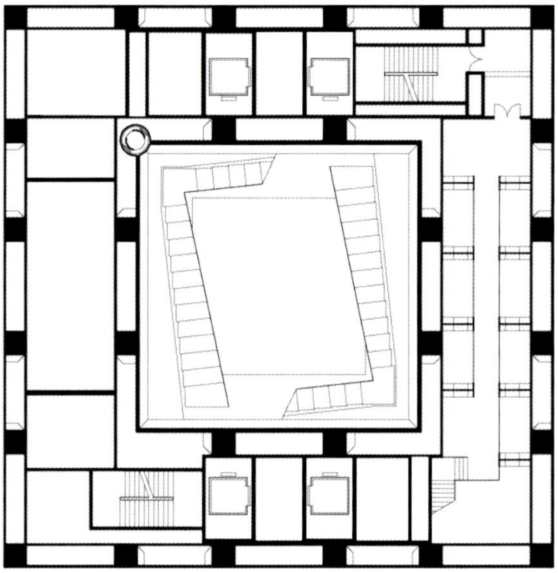

level 6 /

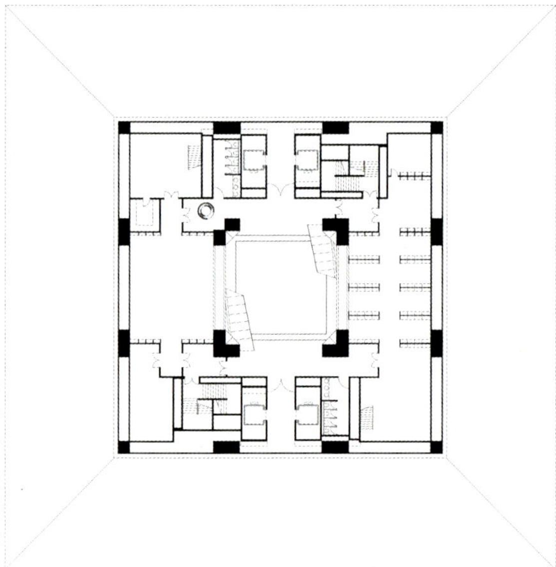

level 5 /

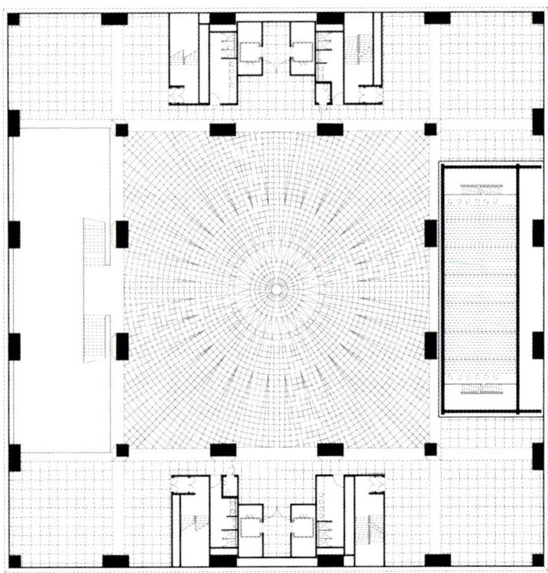

\ level 0

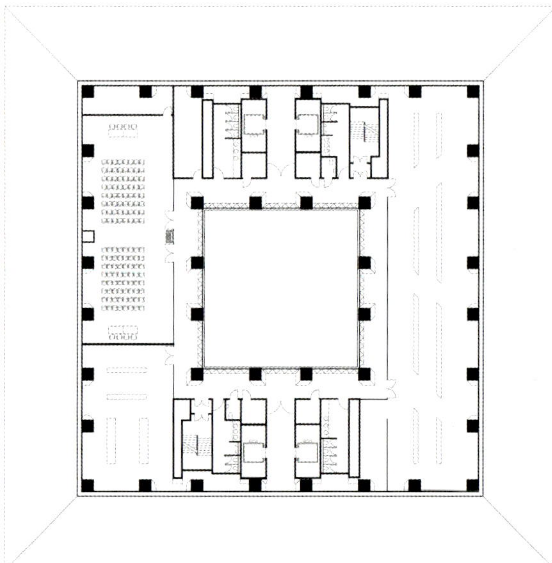

level 3 /

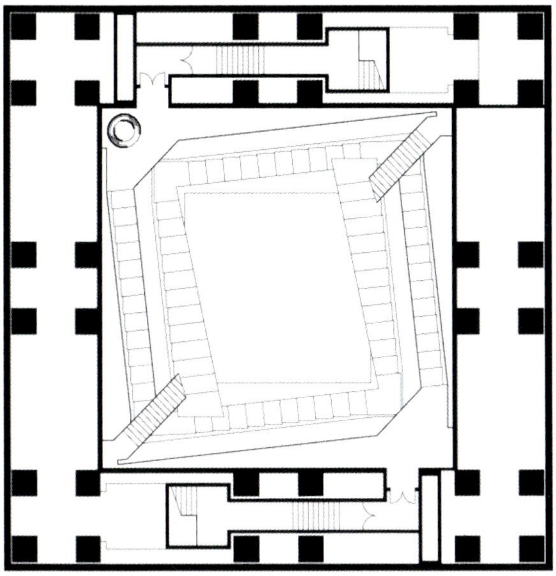

\ level 7

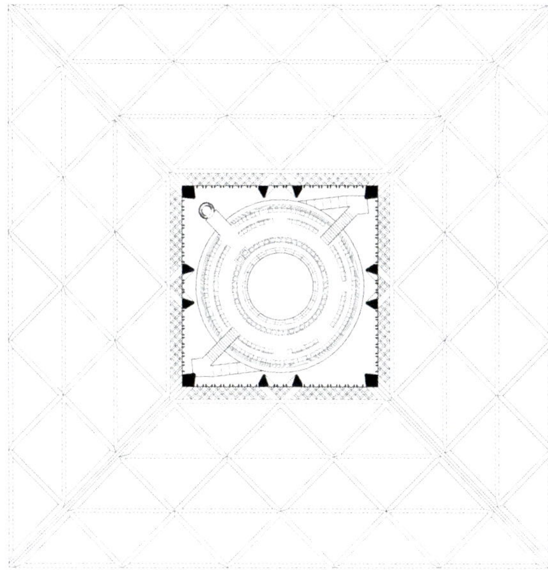

level 8 /

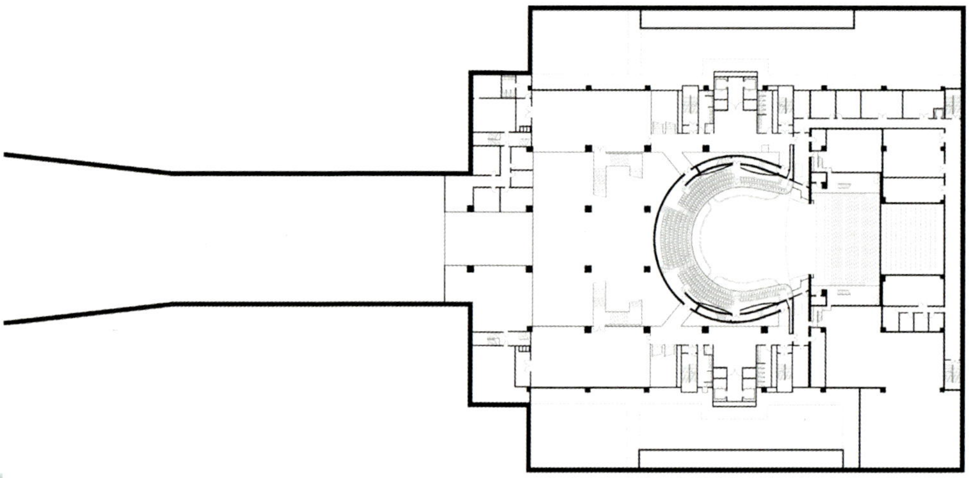

level -1 /

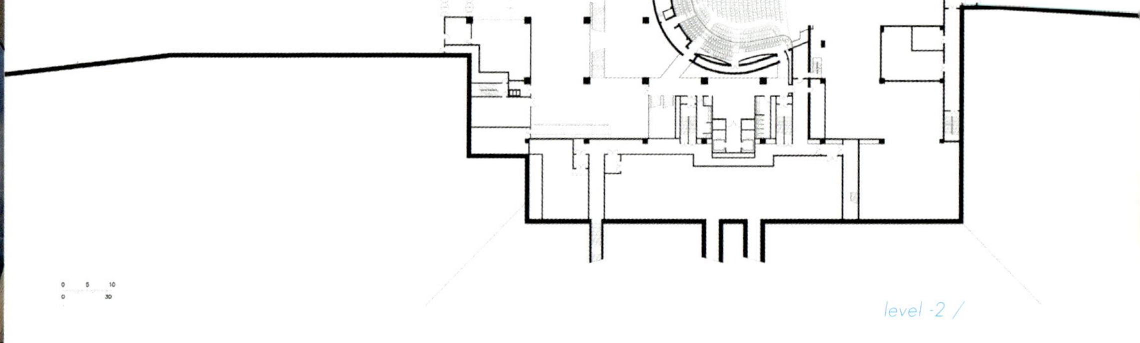

level -2 /

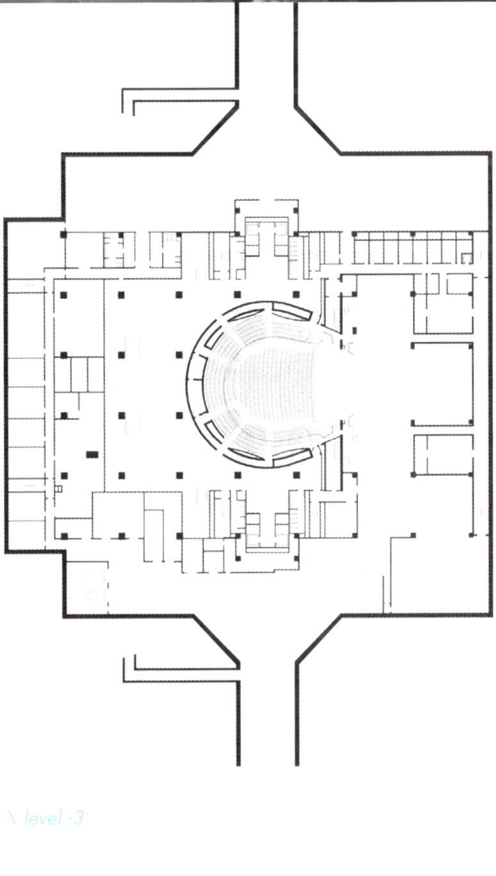

\ level -3

INTEGRATED REVITALIZATION OF URBAN TISSUE

도시의 구조 안에서 변화하는 사회 환경과 프로그램으로 지속적으로 중첩되는 과거의 건축과 새로운 제안 사이의 관계는 동일한 도시의 역사나 기억의 흐름 안에 존재하는 연결성으로 이해할 수 있다. 그 과정에서 표현된 서로 다른 재료와 기술 사이의 몽타주, 제거와 덧붙임으로의 연결로 대비되는 차이는 '역사적 건축에 대한 부정적 판단'으로 볼 수는 없다. 도시의 풍경을 전면으로 드러내며 거리의 파사드에 적용된 새로운 시도와 결과물은 도시에 새로운 비전을 제시하면서 스스로 의도한 건축적 개념을 드러내고자 한다.

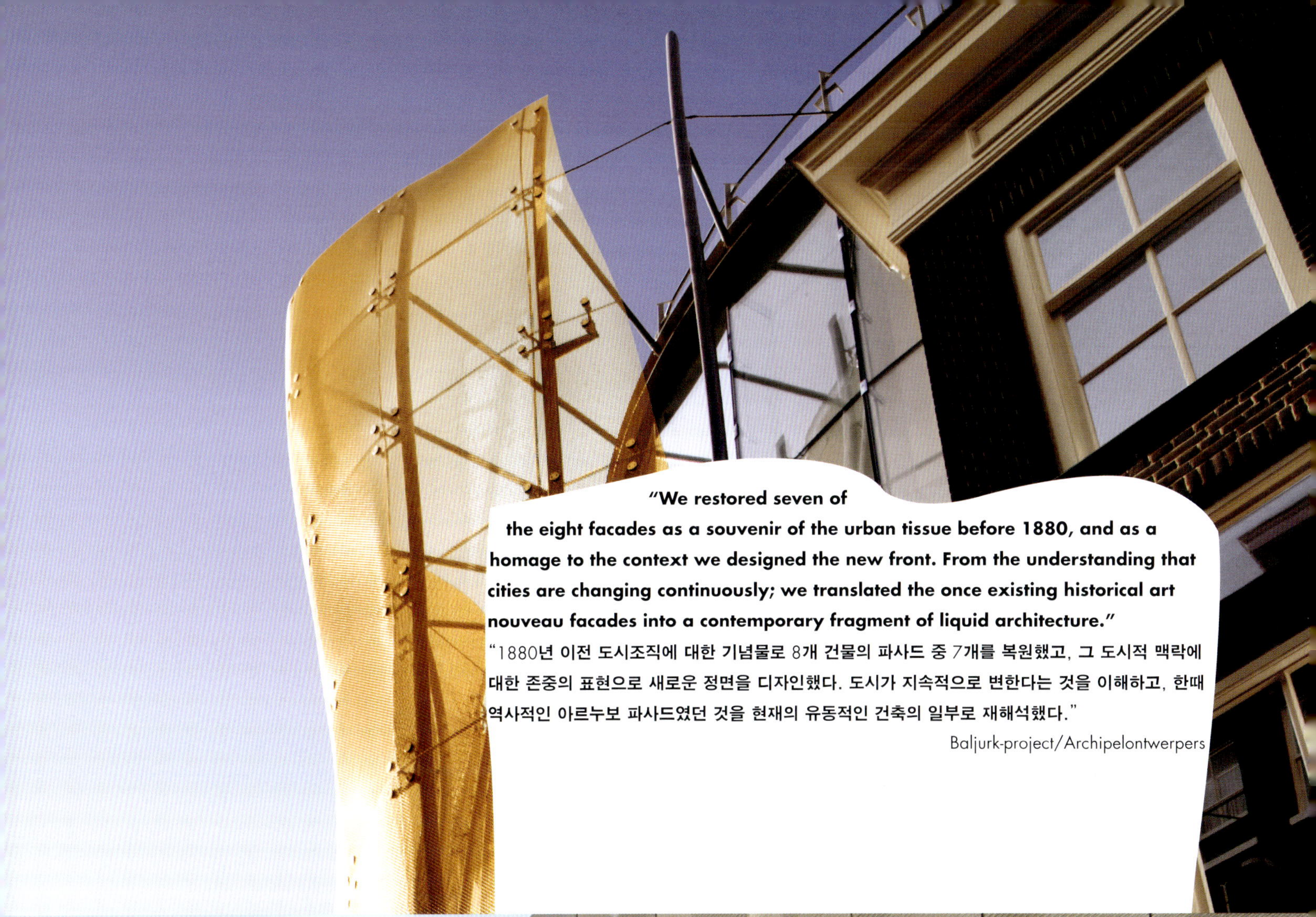

"We restored seven of the eight facades as a souvenir of the urban tissue before 1880, and as a homage to the context we designed the new front. From the understanding that cities are changing continuously; we translated the once existing historical art nouveau facades into a contemporary fragment of liquid architecture."

"1880년 이전 도시조직에 대한 기념물로 8개 건물의 파사드 중 7개를 복원했고, 그 도시적 맥락에 대한 존중의 표현으로 새로운 정면을 디자인했다. 도시가 지속적으로 변한다는 것을 이해하고, 한때 역사적인 아르누보 파사드였던 것을 현재의 유동적인 건축의 일부로 재해석했다."

Baljurk-project/Archipelontwerpers

"The articulation of the facade is generated through a continuous variation and modulation of the vertical tectonics of the facade of the old factory: bending vertical lines in a complex pattern that produce a whole range of effects when walking or driving by, enhanced by the position of the sun."

"파사드의 분절은 연속적인 변화와 옛 공장 파사드의 수직적으로 구축된 것을 변화시킴으로써 이루어진다. 복잡한 패턴 속의 휘어진 수직선은 걸어가거나 차를 타고 지나갈 때 온갖 효과를 만들어내며 태양의 위치에 의해 더욱 풍부해진다."

Maison Folie/NOX

Hague, the Netherlands

2005

Archipelontwerpers

Function_ House, Shopping space

Project architects_ Eric Vreedenburgh, Jaap Baselmans, Coen Bouwmeester, Arjan Mulder

Photographer_ Archipelontwerpers

BALJURK-PROJECT

발주크 프로젝트(Baljurk = evening-dress)는 옛 파사쥬 옆, 헤이그의 중심부에 위치하고 있다. 1882년 파사쥬의 건설은 도시의 조직에 변화를 가져왔다. 이런 변화가 일어나기 전까지 케팅스트랏(Kettingstraat)과 앳테롬(Achterom) 두 거리 사이에는 방향성 없는 변화가 일어나곤 했는데, '엣테롬'이 평범한 개울이었기 때문이다. 이런 유기적인 흐름은 헤이그의 직사각형으로 구획된 환경 속에서 특이한 요소가 된다. 파사쥬가 건설된 결과로 이런 자연스러운 변화는 평각으로 전환된다. 발주크 프로젝트는 바로 이 모퉁이에 놓여있다. 도시 내에서 이 부분에는 다른 모든 역사적인 장소들과 마찬가지로 각기 다른 건축 양식과 유형의 이질적인 요소들이 모여 있다. 네덜란드 켈빈주의 문화에 있어서 상당히 특이한 점으로서 주변에 몇몇 아르누보 양식의 건물들이 있다는 것이다. 지난 20년 동안 이 장소의 도시계획적 기회는 무시되었다. 거리와 골목은 헤이그 중심부에서 황폐한 곳 중의 하나가 되었다. 거기에는 막히거나 쇠창살이 설치된 창문, 그라피티가 그려진 역사적인 벽면, 잡초와 오줌이 뒤섞인 냄새 등이 있다. 다시 말하자면 이곳은 네덜란드 왕실 주거의 '후미진 곳'인 셈이다.

발주크를 위한 계획은 이 주변 환경에 대한 종합적인 재생의 한 요소였고, 우리의 계획에는 8개의 역사적인 장소가 포함된다. 이러한 파사드의 뒤 들어 올린 레벨에 놓인 특별한 주택 8채를 설계했다. 이 레벨의 하부에는 쇼핑을 위한 공간을 계획되었다. 1880년 이전 도시조직에 대한 기념품으로서 8개 건물의 파사드 중 7개를 복원했고, 그 도시적 맥락에 대한 존중의 표현으로 새로운 정면을 디자인했다. 도시가 지속적으로 변한다는 것을 이해하고, 한때 역사적인 아르누보 파사드였던 것을 현재의 유동적인 건축의 일부로 재해석했다. 발주크의 파사드를 구성하기 위해서 더블스킨을 적용했다. 내부골격은 유리 커튼월로 지어졌고, 이 정면은 실제적인 외벽의 역할을 한다. 외부골격은 철골로 물결치듯 짜인 보일(voile)이다. 이것은 햇빛을 여과시키지만 이보다 더 중요한 것은 이 정면부가 이곳의 '새로운 시대'를 재현하고 있다는 점이다. 이 두 표피 사이에는 철재 튜브와 텐션 로드(tension rods)같은 것으로 구성된 섬세한 우산을 디자인했다. 거미는 철재의 직물과 하부구조물을 통합한다. '골든 스크린'은 스테인리스 스틸로 만들어졌고 특별한 색채를 만들어내기 위해 특수한 '크롬(chrome)'으로 처리된 것이다. 제조업자는 이 황금빛 청동의 색채가 50년이 넘도록 지속적으로 빛날 것임을 보장했다.

The 'Baljurk-project'(baljurk = evening-dress) is situated in the centre of The Hague, next to the old Passage. The construction of the Passage in 1882 has transformed the urban tissue. Before this transformation there used to be a meandering transition between the two streets 'Kettingstraat' and the 'Achterom', caused by the fact that the 'Achterom' used to be a modest Project architects_ Eric Vreedenburgh, Jaap Baselmans, Coen Bouwmeester, Arjan Mulder
. This organic flow is unique in the context of The Hague's rectangular lay out. By consequence of the construction of the Passage, this smooth transition was converted into a straight angle. The Baljurk-project is exactly placed on this corner.

This part of the city (like every historical part of the city) contains a heterogeneous collection of different architectural styles and typologies. Quite unique (for a Calvinistic culture like the Dutch one) are the presence of some art nouveau buildings in the environment.

For the last two decades the urban opportunities of this place has not been noticed. The streets and alleys became one of the most desolate parts of the centre of The Hague: blind and armoured windows, graffiti on the historical walls and the smell of a fusion bouquet of weed and urine. In a word; this was 'the backside' of the Dutch royal residence. The plan for the Baljurk is a component of the integrated revitalisation of this neighbourhood. Our plan contains eight historical plots. Behind these facades we designed eight special houses, build on an elevated level. Under this level we planned shopping space.

We restored seven of the eight facades as a souvenir of the urban tissue before 1880, and as a homage to the context we designed the new front. From the understanding that cities are changing continuously; we translated the once existing historical art nouveau facades into a contemporary fragment of liquid architecture.

We drafted a double skin to compose the facade of the Baljurk. The inner shell is build up as a curtain wall of glass, this front works as the physical skin. The outer shell is the woven wave of steel: the voile. It filters the sunlight, but more important it represents the 'new age' of this neighbourhood.

Between the two skins we designed a fragile umbrella like construction made of steel tubes and tension rods. Spiders join the woven steel with its sub-construction. The 'golden screen' is made of stainless steel and is treated with a special 'chrome' to obtain this special colour. The manufacturer guaranteed us that this golden bronze colour will shine and twinkle more than fifty years steady.

Archipelontwerpers

\ site plan

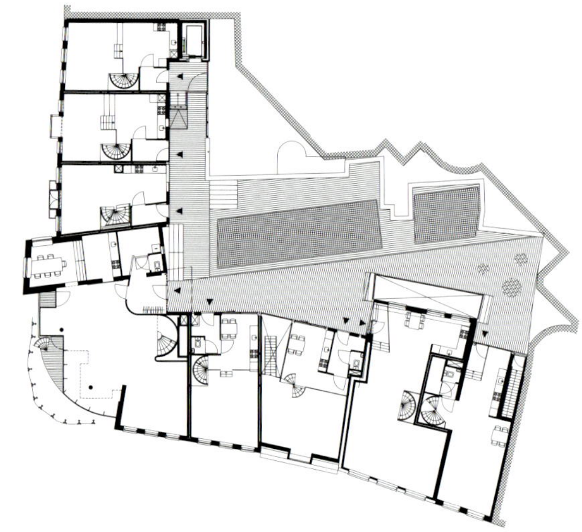

2nd floor plan /

1st floor plan /

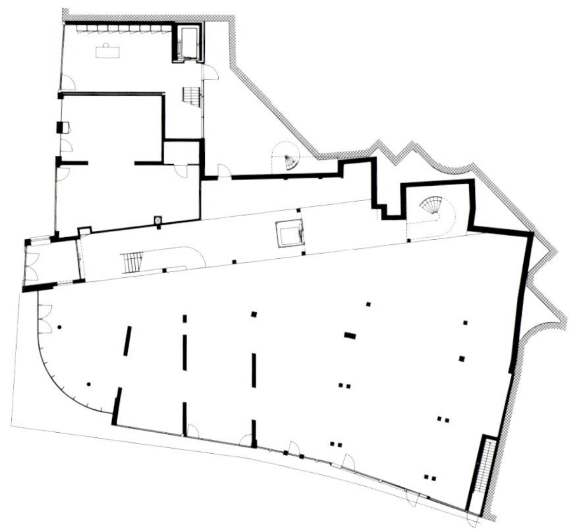

ground floor plan /

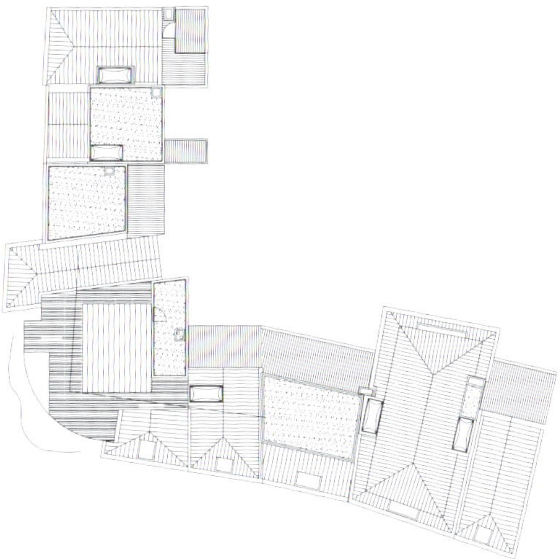

\ 5th floor plan

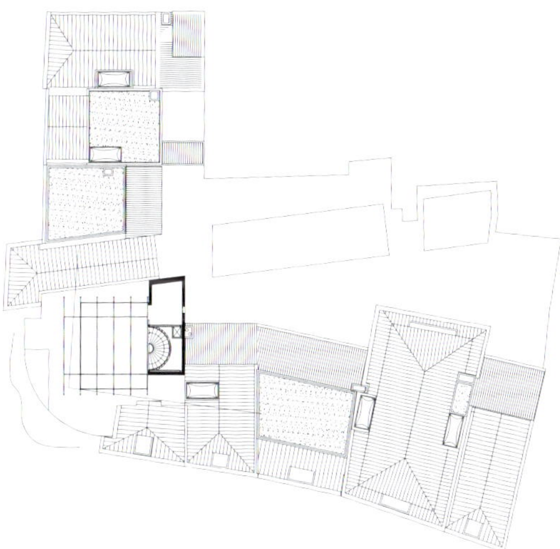

\ 4th floor plan

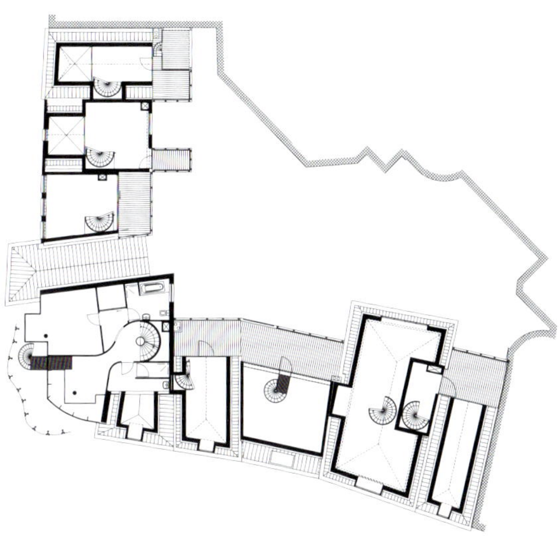

\ 3rd floor plan

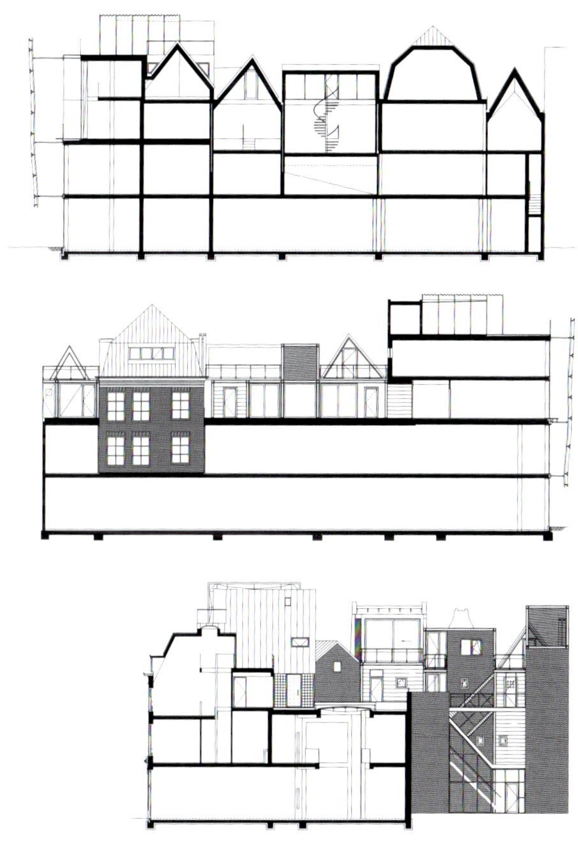

sections /

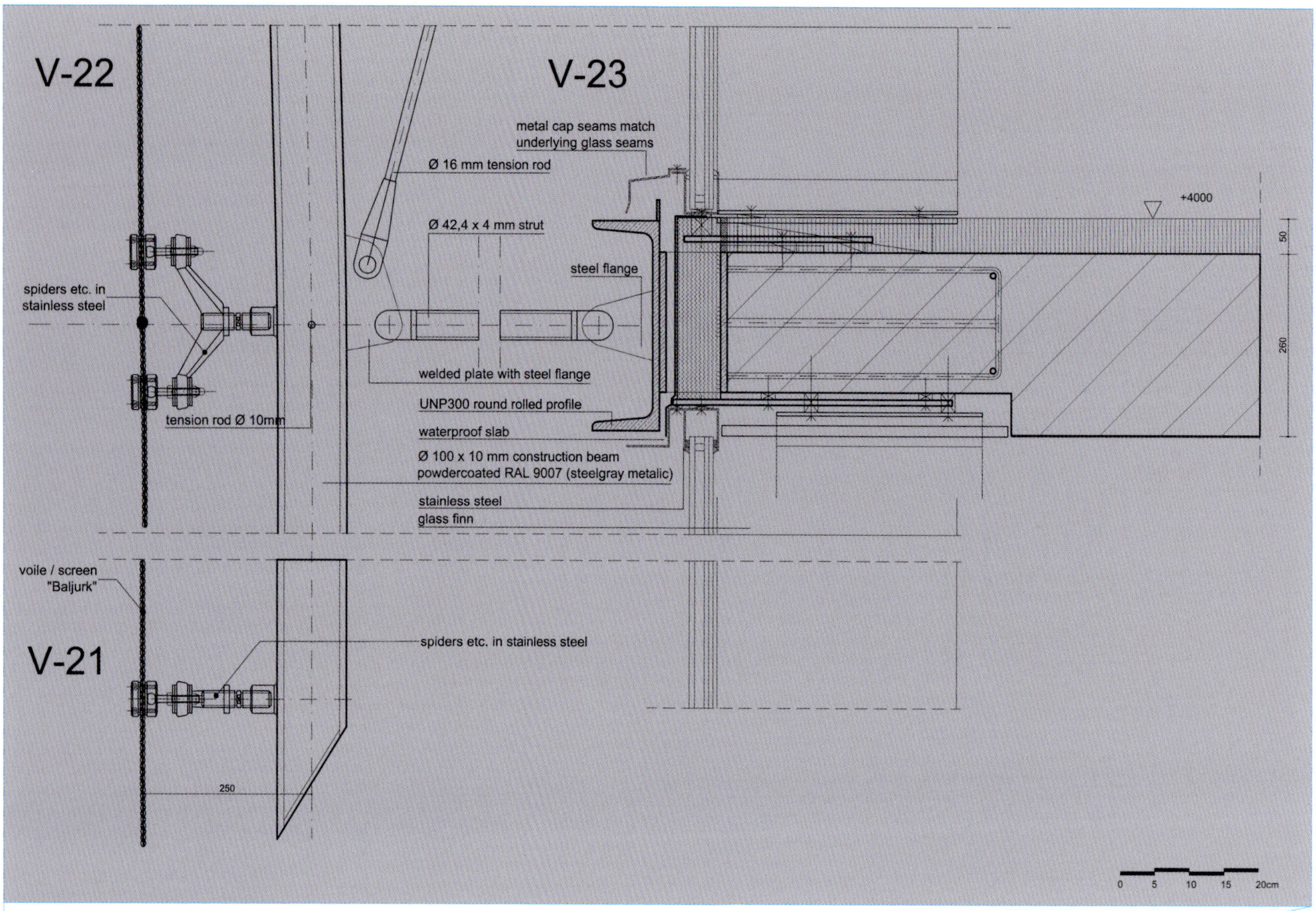

screen facade detail a

screen facade detail b

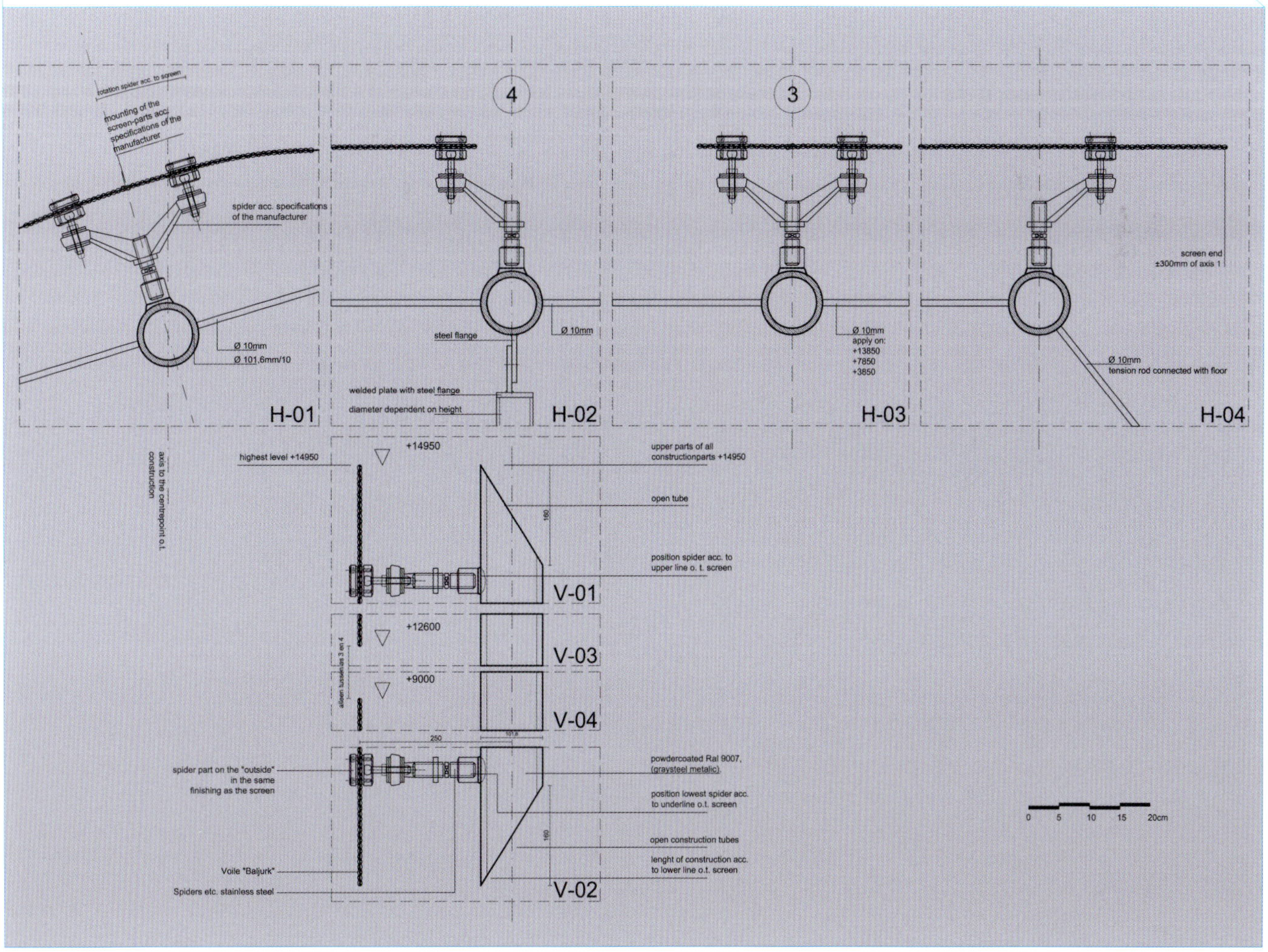

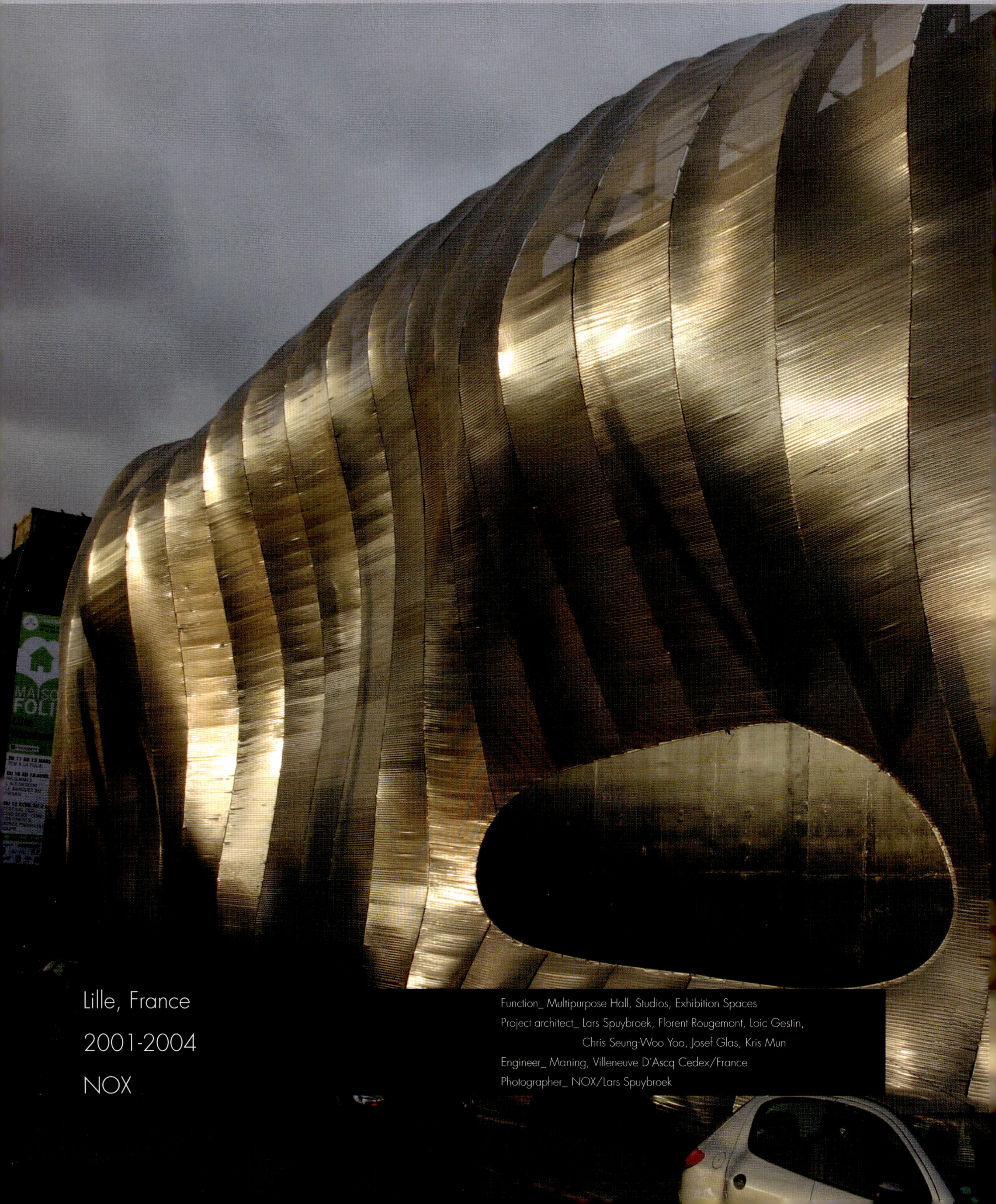

Lille, France

2001-2004

NOX

Function_ Multipurpose Hall, Studios, Exhibition Spaces

Project architect_ Lars Spuybroek, Florent Rougemont, Loic Gestin, Chris Seung-Woo Yoo, Josef Glas, Kris Mun

Engineer_ Maning, Villeneuve D'Ascq Cedex/France

Photographer_ NOX/Lars Spuybroek

MAISON FOLIE

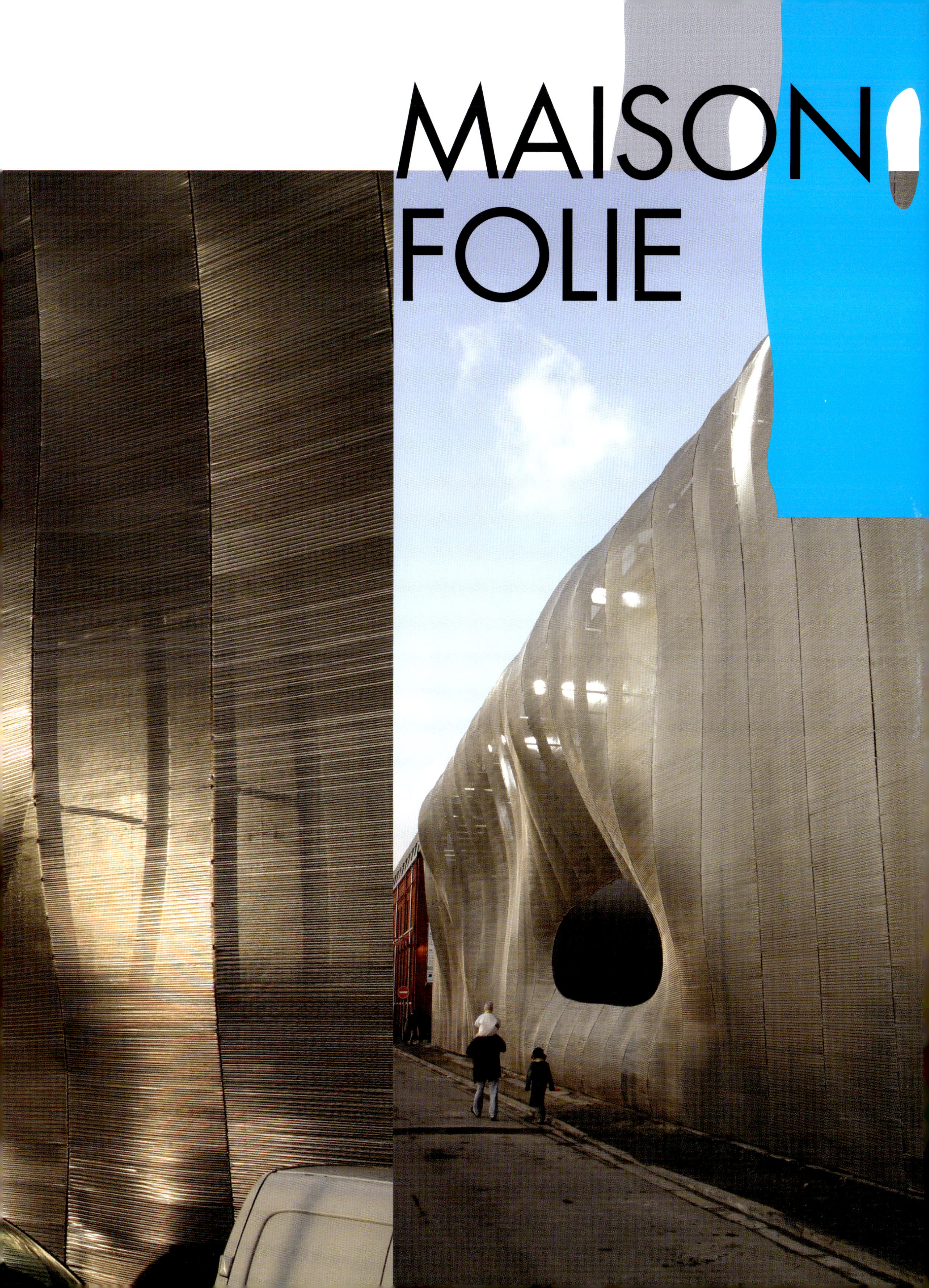

이 건물군은 오래된 직물공장을 예술과 관련된 기능과 콘서트, 연극 제작, 패션쇼 같은 용도로 사용될 새롭게 건설된 다목적 홀이 더해진 복합적인 건물로 리노베이션한 것이다. 로테르담, 함부르크, 혹은 글래스고우(Glasgow)에서 예술 활동을 위한 무대는 버려진 산업용 건물에 수용되어 도시의 정체성을 변화시키고 다시 되살아나게 한다. 도시로 만들어진 도시인 릴(Lille)에서 네트워크에 대한 개념이 유행하고 있다. 그것은 단지 생존을 위해서가 아니라, 생명력의 원천이기도 하다. 근본적으로 이 생각에는 규모의 제한이 없다. 즉, 네트워크가 네트워크를 품고 있는 것이다. 릴은 초고속 글로벌 네트워크 안에서 번성하고 있지만, 일 그 자체 또한 하나의 네트워크, 즉 다결절도시이다. 메종 폴리(Maison Folie)에 대한 아이디어는 이러한 결절점들 중의 하나인 와제메스(Wazemmes) 속에 있는 작고 더 정교한 네트워크다. 와제메스의 메종 폴리에 대한 아이디어는 세 가지 각각 다른 차원에서, 이런 네트워크에 대한 아이디어에 기초를 두고 있다.

지역사회의 맥락 안에서 이 프로그램은 와제메스 지구 내에 활기차고 생생한 교류를 촉진한다. 그것은 특별히 오래된 공장 안에 적절히 자리를 잡고, 작은 도시 같은 규모의 차별화로 만들어 진 것이다. 그러므로 간소함은 설계상의 키워드가 된다. 더 경제적인 것이 더 많은 프로그램을 제공할 수 있다. 양적으로는 기존 건물들을 존중하면서 간소하게 처리할 것이 요구되고, 질적으로는 내부적인 다양성(연령, 문화 등)과 함께 대지에 대한 매우 정교한 해결책이 요구된다.

프로그램의 구성요소인 사회적 교류는 끊임없이 연결된 표피, 즉 연속성을 띤 건축물에 의해 촉진된다. 터키탕과 어린이놀이터, 레스토랑, 예술학교, 전시회장, 파티장, 예술가의 집, 영화관, 이웃한 주거지가 서로 연결된다. 하루 중 각각 다른 시각에 따라, 한 주의 요일에 따라 진행되는 프로그램에 의한 다양한 변화가 있으리라고 어렵지 않게 상상할 수 있다. 이 프로그램은 우연한 만남을 촉진하는 건축물이 필요하다. 이런 모든 기능은 개인정원과 공공정원 등, 소규모 녹색 조각 정원을 포함한 하나의 물결치는 조경과 광물 정원을 통해 통합된다. 조경은 모든 요소들이 더 원활한 상호작용을 만들어내도록 하면서 활기에 활기를 더해주고 있다.

신축건물(영화관)에 요구되는 프로그램은 매우 기술적이며 내향적이다. 기본적으로 그것은 도시(현관홀 제외)로부터 차단된 상자들과 와제메스의 주변 도시계획에 꼭 맞는 앙상블을 이루며 최대의 효과를 얻는다. 특히, 와제메스는 심각하게 훼손된 지역이기 때문에 거리를 파사드로 구성하여 마무리하기로 결정되었다. 발광 표피로 이루어진 검은 상자가 외부로 빛을 발하고, 내부의 움직임과 매종 폴리 주변의 움직임과 함께 변화한다. 이 빛나는, 아니 거의 홀로그램에 가까운 외관은 모든 예술과 생활의 파동을 통합한다. 파사드의 분절은 연속적인 변화와 옛 공장 파사드의 수직적으로 구축된 것을 변화시킴으로써 이루어진다. 복잡한 패턴 속의 휘어진 수직선은 걸어가거나 차를 타고 지나갈 때 온갖 효과를 만들어내며 태양의 위치에 의해 더욱 풍부해진다.

A complex of buildings that consists of the renovation of an old textile factory into a conglomerate of art-related functions plus a newly constructed multi-purpose hall, that is used for concerts, theatre productions, fashion shows and the like. As in Rotterdam, Hamburg or Glasgow art moves into disused industrial buildings to transform and revive the identity of a city.

In Lille, a city made of cities, the idea of the network prevails. It is not just necessary to survive, it is also a basis for vitality. Fundamentally this idea is scaleless: networks contain networks. Lille florishes within a global network of high speeds, but Lille is in itself also a network, a multinodal city. The idea of the Maison Folie is again a smaller and even finer network within one of these nodes, Wazemmes. The design for the Maison Folie de Wazemmes is based, on three different levels, on this idea of the network:

In its social context the program stimulates a lively and vital crisscrossing within the district of Wazemmes, it's especially in its fitting within the old factory and its differentiation of scales, which is more like a small city. Austerity is therefore the keyword for this part of the design: the more economical we are, the more program we will be able to provide for. Quantatively it requires a respectful and modest treatment of the existing buildings and qualitatively the program, with its internal dynamics of differences (age, culture) needs a very precise treatment of the ground. The social exchanges, the interaction between programmatic elements is timulated by an architecture of continuity, an uninterrupted connective surface. Connecting hammam to children's playground, to brasserie, to artschool, to exposition, to partyroom, to the home of the artist, to the salle de spectacle, to the inhabitants of the neighborhood. It is not difficult to imagine the different fluctuations of the program at different times of the day, at different days of the week. The program needs an architecture that stimulates unexpected encounters. We have incorporated all these functions within one undulating landscape, a mineral garden, that includes patches of smaller green gardens, either private or public. The landscape makes everything interact more easily, and feeds activity to activity.

The section of the program that requires a new building (salle de spectacle with its studios) is both a very technical but also introverted program: basically it is best served by an assemblage of boxes that shut themselves off the city (except for the foyer) but fit exactly within the urban scheme of the Wazemmes neighborhood. Especially since Wazemmes is such a damaged area we decided to finish the street with a facade. We made the black box glow externally with a luminous skin which transforms with movement in and around the Maison Folie. This glowing, almost holographic dress incorporates all the pulsations of art and life. The articulation of the facade is generated through a continuous variation and modulation of the vertical tectonics of the facade of the old factory: bending vertical lines in a complex pattern that produce a whole range of effects when walking or driving by, enhanced by the position of the sun.

NOX

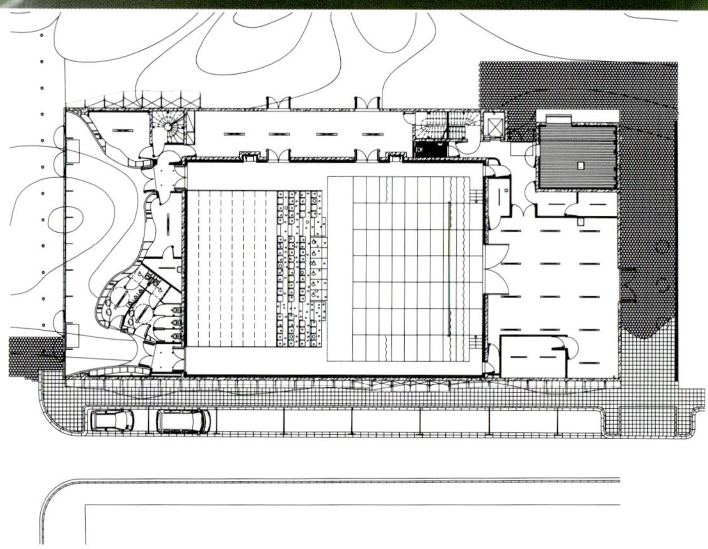

\ level 0

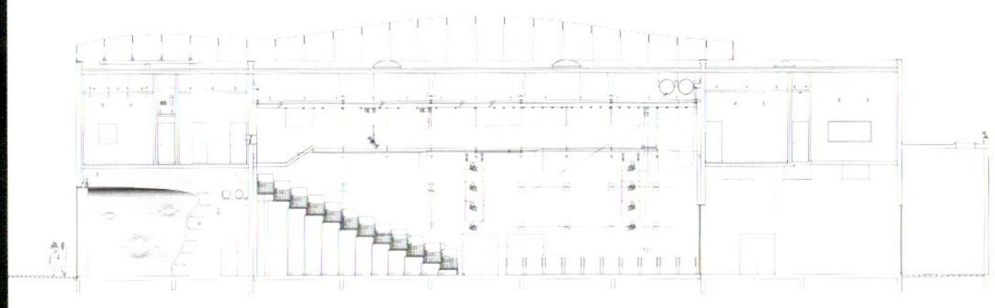

longitudinal section /

facade /

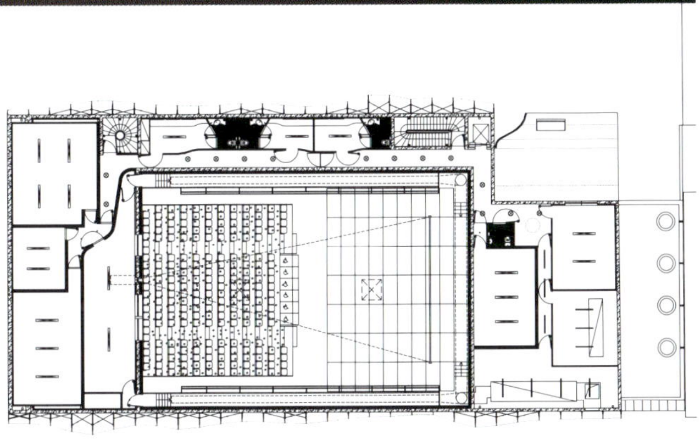

\ level 1

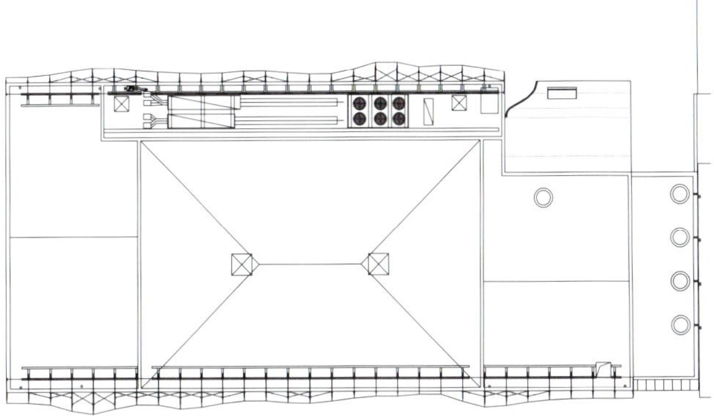

\ level 2

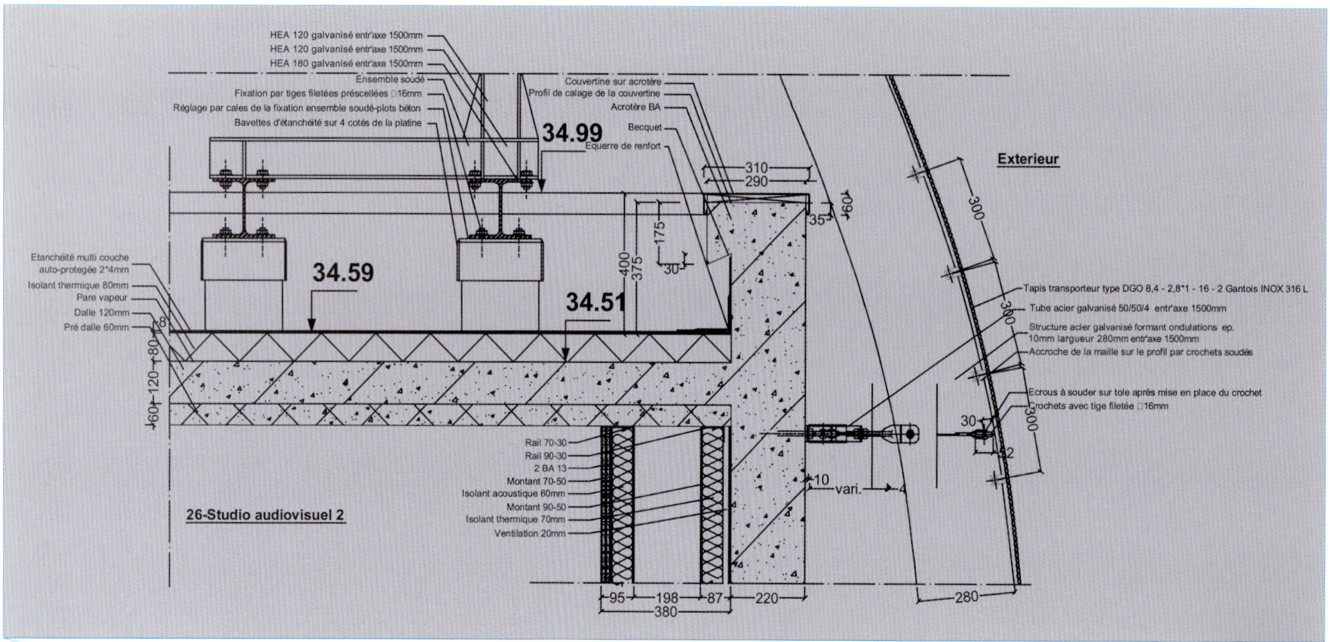

detail AA

detail BB

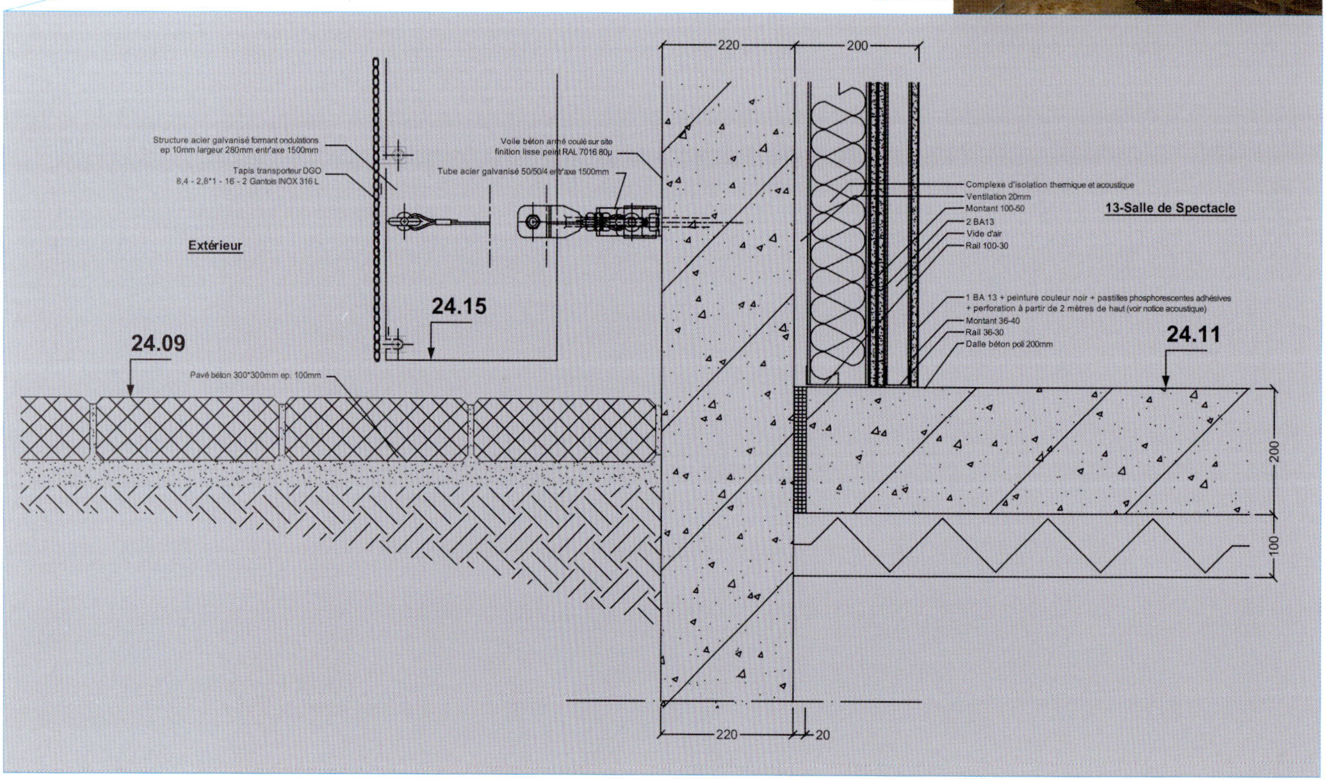

facade details

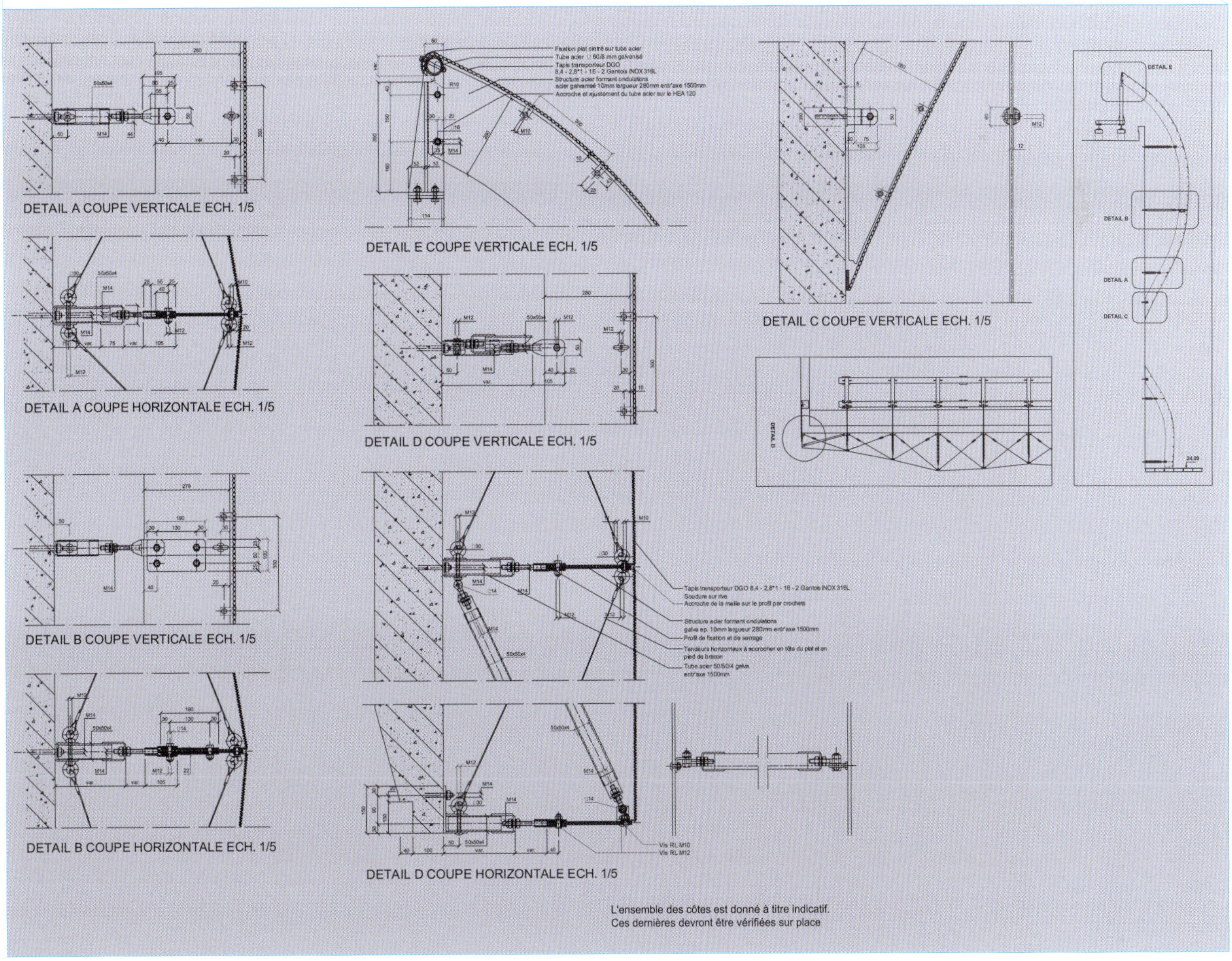

Sydney, Austrailia

2003-2005

Bligh voller nield

Function_ Commercial building

Total floor area_ 10,000m^2

Project architect_ Bligh voller nield

Structural Engineer_ Sinclair Knight Merz

Photographer_ Bligh voller nield

Client_ DB Real Estate (on behalf of SAS Trustee Corporation)

NORTEL NETWORKS HEAD OFFICE

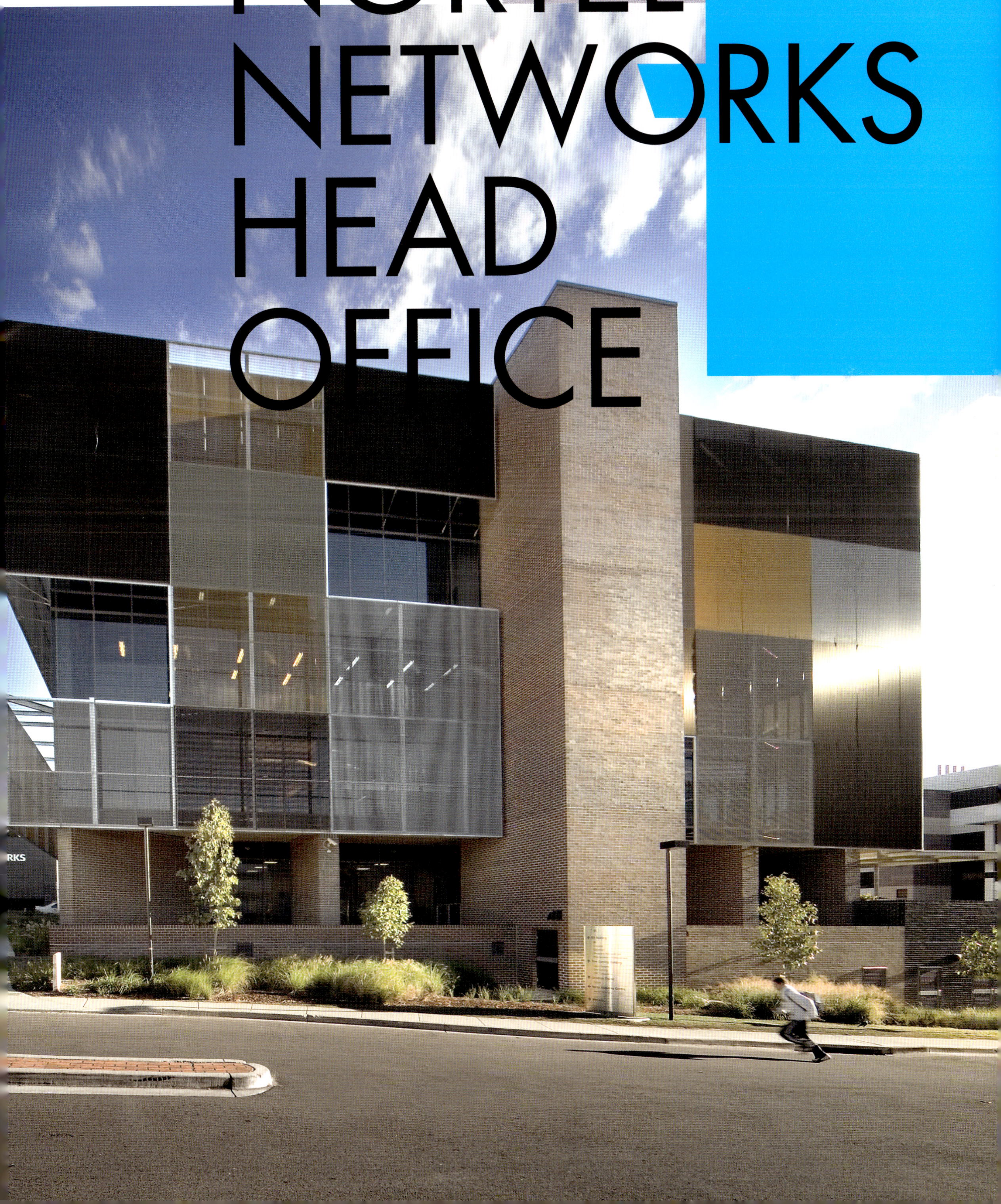

매커리(Macquarie) 대학교 리서치 파크 내의 노스 라이드(North Ryde), 이노베이션(Innovation) 가 1번지에 위치한 5층 규모의 사무실 건물은 지하 주차장과 임대 가능한 사무실 공간 10,000m²를 제공한다.

사무실 층은 중앙의 아트리움을 중심으로 형성되었는데, 각 층당 1,000m² 면적의 두 바닥층을 형성한다. 노텔 네트웍스(Nortel Networks)는 주요 입주자로서 6,500m²를 사용하고 있다.

아트리움 공간은 각 층의 사무실로 자연광을 공급하는 커다란 천창까지 6개 층에 걸쳐 형성되었다. 엘리베이터는 개방된 로비가 있는 아트리움 내부에서 각 층을 연결한다.

건물의 외벽은 층 사이 전체를 채광하는 검은색 '유리판넬(Vitrepanel)'클래딩이 스팬드럴(spandrels)과 파라펫(parapet)에 적용된 알루미늄 프레임으로 구성된다. 이런 어두운 배경에 대응해 알루미늄 블레이드 스크린이 건물의 외벽에서 1.2미터 떨어진 아연도금 철제 프레임 위에 적용되었고, 사무실 층과 아트리움의 자연채광 조절을 돕는다.

외부 건축 공간 구성과 내부 환경의 다양성을 확보할 기회를 마련하기 위해, 판넬 블레이드는 수평 수직적으로 다양하게 배열되었다. 몇몇 판넬의 조합들은 뒷면의 유리 파사드를 드러내며 생략되기도 한다. 스크린은 수평 판넬 및 색을 입힌 부분과 빈 공간을 조합하는 프레임을 구축하는 4.2미터 X 4.8미터의 기본 모듈로 제작되었다.

The five storey office building at 1 Innovation Road, North Ryde in the Macquarie University Research Park provides 10,000m^2 of nett lettable office area with car parking in the basement levels.

The office floors are set around a central atrium forming two 1,000m^2 floor plates at each floor. Nortel Networks, the major tenant occupies 6,500m^2.

The atrium space rises through 6 storeys with extensive roof glazing allowing daylight to penetrate to the office floors at each level. The lifts are positioned in the atrium space with open lobby landings and bridges at each floor.

The building enclosure comprises floor to floor full height glazing in dark grey aluminium framing with black 'Vitrepanel' cladding to the spandrels and parapet. Against this dark backdrop, aluminium blade screens on galvanised steel frames 1.2 metres from the face of the building assist in the control of daylight glare on the office floors and atrium.

For external architectural composition and to provide the opportunity for a variety of internal environments, the arrangement of blades on the panels varies between horizontal and vertical. Some groups of panels are omitted revealing the glass facade behind. The screens are made up of typical 4.2 metre x 4.8 metre modules establishing a compositional frame for the horizontal and vertical panel, colours and voids.

Bligh voller nield

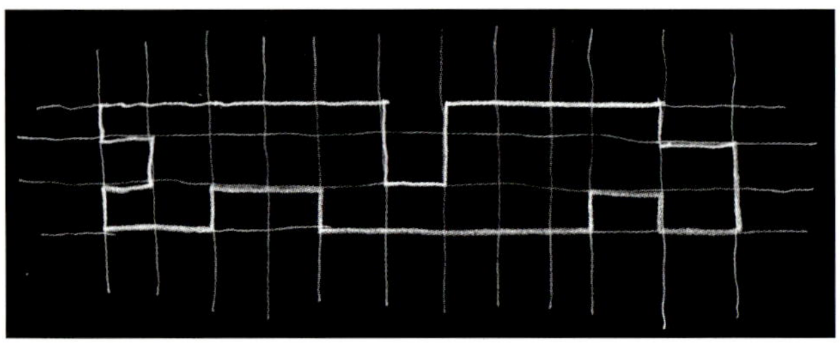

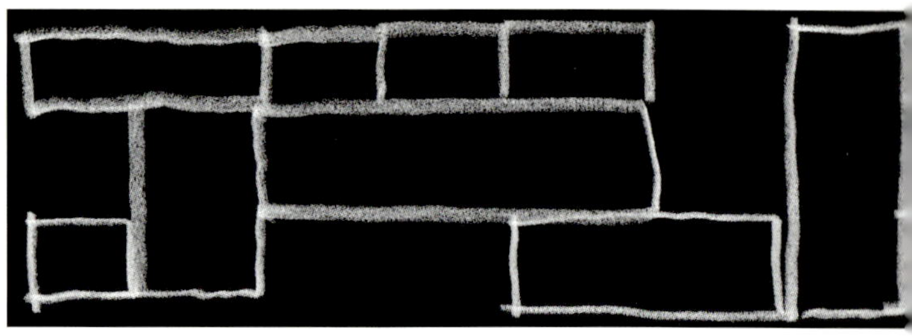

\ *sketch*

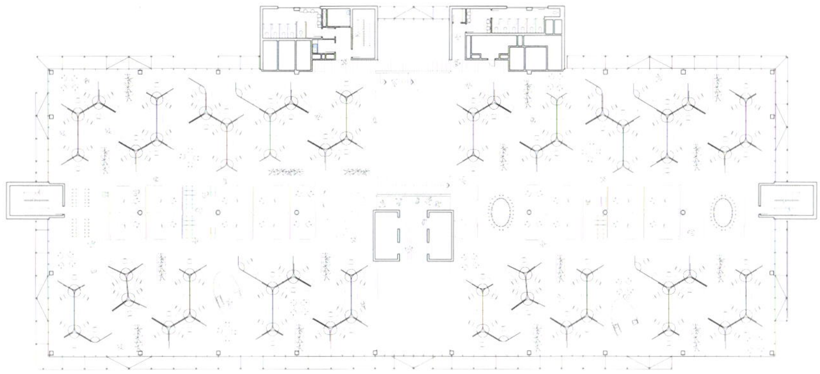

typical floor plan /

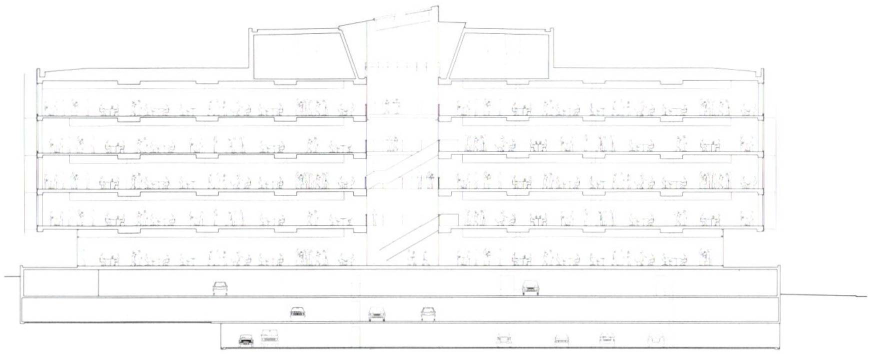

section /

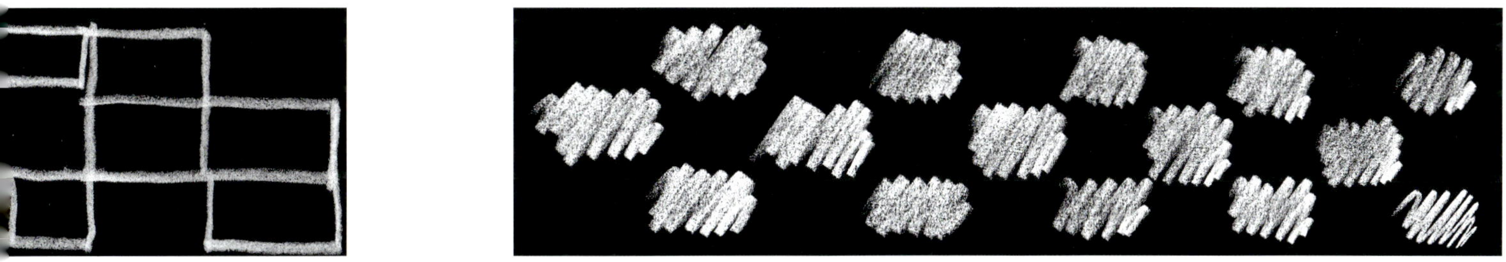

concept of facade /

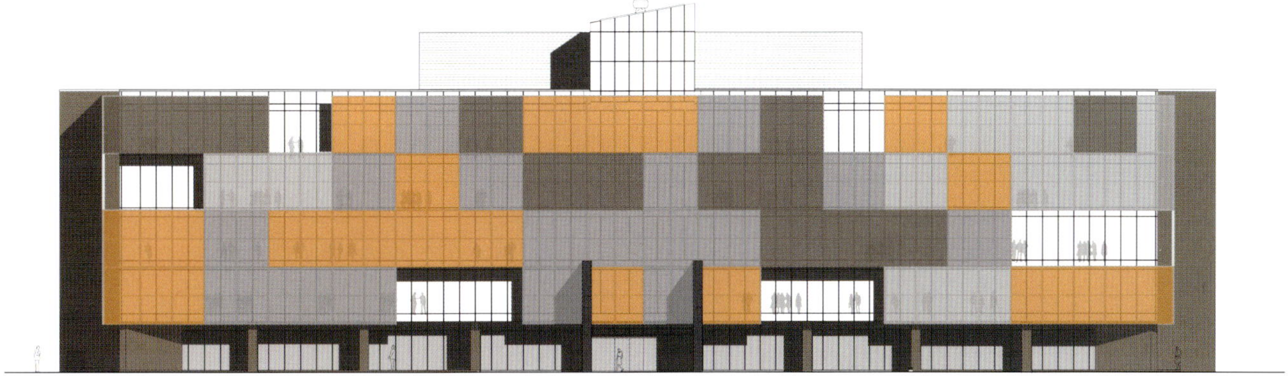

\south east elevation

NORTEL
NETW

facade & screen detail

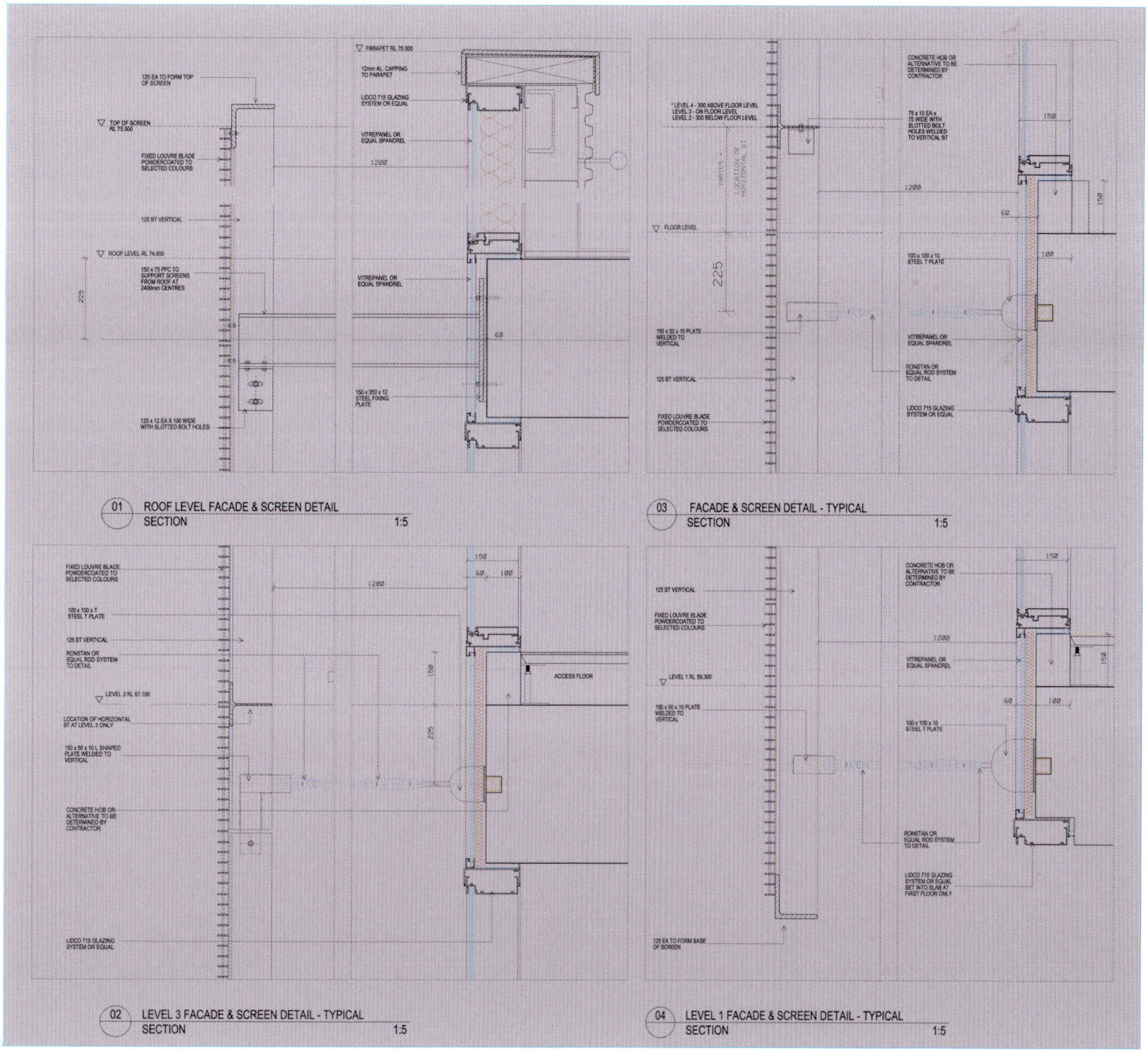

NEIGHBOR OFFICE BUILDINGS IN THE ZILVERPARKKADE

West 8에 의해 설계된 렐리스타트(Lelystad)시의 질베르파케이드(Zilverparkkade)에는 밀집한 도심에 대한 도시계획에 따라 구획된 사무실 용도로 사용될 12개의 부지가 묶인 클러스터가 있다. 여기에 에릭 반 에거라트(Erick van Egeraat)건축사무소, 르네 반 주크(Rene van Zuuk)건축사무소, 텟카(Tekta) 건축사무소가 디자인한 건물이 나란히 들어섰다. 이 건물들은 17세기 네덜란드의 운하주택을 바라보며 좁은 대지에 줄지어 서 있다. 엄격히 구획된 대지의 조건 안에서 차별성을 만들어낼 수 있는 요소는 '파사드'가 유일하다. 여기에서는 비슷한 시기에 완공된 'Zilverparkkade D by Rene van Zuuk Architekten'와 'Office Zilverparkkade by Erick van Egeraat'를 통하여 동일한 대지의 조건 위에서 각기 다른 시각을 가진 두 건축가가 파사드를 통해 표현한 개성 넘치는 아이디어를 살펴보고자 한다.

"The works of the Dutch graphic artist Maurits Escher have been an unmistakable model. Specific skills are required to devise a repetitive pattern that, applied in a limited number of different concrete elements, constitutes a seamless entity. The craftsmanship needed to avoid the repeating units to be too obvious, has an analogy with designing rotation press patterns for textile and wallpaper."

"네덜란드의 그래픽 아티스트 모리츠 에셔(Maurits Escher)의 작품이 그 모델이 되었는데, 한정된 종류의 콘크리트 요소를 적용해 단절 없이 하나로 완결하는 반복적인 패턴을 고안하려면 특별한 기술이 요구된다. 이런 반복적인 요소들이 눈에 잘 띄지 않도록 하는 것이 장인정신이며, 이러한 반복적인 요소들은 직물과 벽지를 위한 순환인쇄패턴의 디자인에서 유추되었다."

Zilverparkkade D/ Rene van Zuuk Architekten

"Inspiration for the design of the facade was found in the image of a Venetian palazzo. The stateliness, the composition and ornamentation are translated into a contemporary appearance. The result is a building with a strong architectonic expression, answering the needs and measurements of the location, yet satisfying all the demands that are requested of a modern office building."

"입면 디자인에 대한 첫인상에서 베네치아 팔라조(Venetian palazzo)의 이미지를 발견할 수 있다. 위엄성과 구성 그리고 장식성이 현대적으로 변형되어 나타난다. 이것은 강렬한 건축적 구축의 표현이고, 입지의 규모적 요구에 대한 대답인 동시에 현대적인 사무소 건축물에서 요구되는 모든 필요조건을 만족하고 있다."

Office Zilverparkkade / Erick van Egeraat

Lelystad, the Netherlands

2002-2006

EEA

Function_ Office

Building floor area_ 2,168m²

Project architect_ Erick van Egeraat

Photographer_ Chistian Richters

Client_ Ter Steege Vastgoed

OFFICE ZILVER-PARKKADE

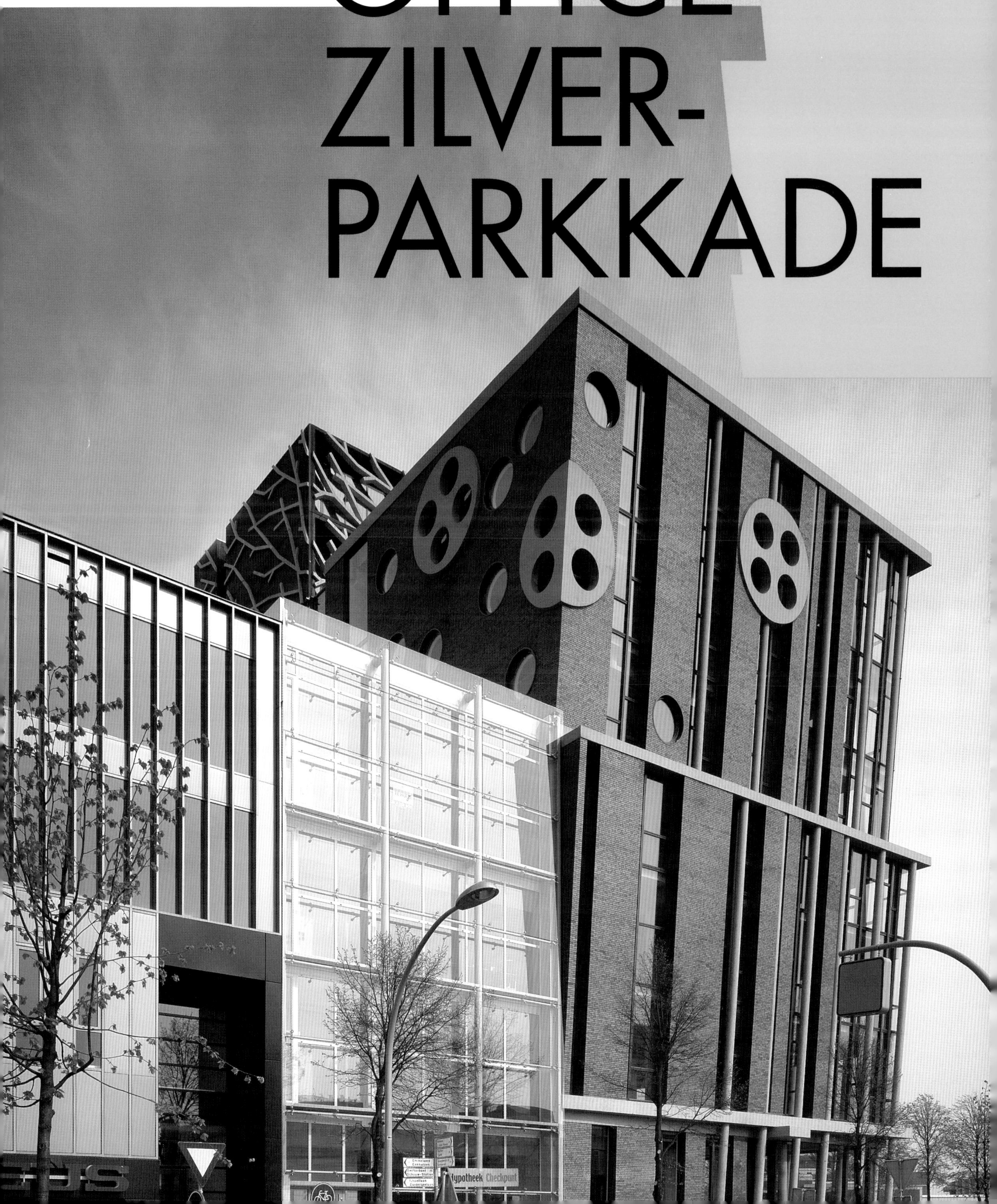

오피스 빌딩인 '질베르파카케이드(Zilverparkkade)'는 네덜란드의 렐리스타트(Lelystad)시 중심 지역에 입지한다. 기차역으로 향하는 주 출입로와 인접한 대지는 도시 주변 지역에서 오피스와 아파트 개발 계획을 하는 도시계획사무소 West 8의 마스터플랜을 토대로 형성되었다. 렐리스타트를 개발한 건축주는 최상의 복합오피스빌딩이 되기를 희망하였다. 임대지역은 2,000m²의 대지면적으로 구성되며, 7층 이상으로 나누어진다. 지상층은 다른 층과는 다르게 5.5미터의 층고를 가진다. 그리고 층 사이에 중층을 두어 지상층으로부터 직접적인 연결이 가능하도록 한다. 건물은 살짝 이동하며 선회하는 세 개의 볼륨으로 구성되는데, 그 형태는 인접한 건물들의 규모와 비슷하게 구성된다.

기둥과 벽돌마감은 건물 전체 높이를 넘어 연장되어 입면의 수직성을 돋보이게 한다. 나무재질의 창문 프레임 역시 수직성을 두드러지게 한다. 보와 기둥 그리고 입면의 장식은 콘크리트의 무게감과 동시에 부드러움을 드러내고 있다. 그것은 경량 유리 알갱이와 자연의 외형을 연출하는 2가지 색상의 회반죽으로 마감되었다. 코너 쪽으로 벽돌은 유리창으로 대체되면서, 코너에 최대의 개방감을 만들어 낸다. 입면 디자인에 대한 첫인상에서 베네치아 팔라조(Venetian palazzo)의 이미지를 발견할 수 있다. 위엄성과 구성 그리고 장식성이 현대적으로 변형되어 나타난다. 이것은 강렬한 건축적 구축의 표현이고, 입지의 규모적 요구에 대한 대답인 동시에 현대적인 사무소 건축물에서 요구되는 모든 필요조건을 만족하고 있다.

The office building 'Zilverparkkade' is built on a prominent site in the city centre of Lelystad, the Netherlands. Located adjacent to the main entry road to the train station, the site forms the closing cornerstone of the master plan by Urban Planners West 8; comprising of a development scheme for offices and apartments at the edge of the city. The client, Develop Lelystad, envisioned a high-quality, multi-use office building. Rental area is comprised of 2.000m^2, which is divided over seven floors. The ground floor has an extraordinary height of 5,5m. An additional entresol is directly accessible from the ground floor. The composition of the building consists of three slightly shifting, stacking volumes, which connect to the scale of the adjacent buildings. Columns and brickwork extend over the full height, thus accentuating the verticality of the elevation. Wooden window frames also accentuate verticality. The beams, columns and ornaments of the facade suggest the smoothness and heaviness of concrete. They are constructed with lightweight glass granulate and finished in a 2 colour blend of smooth plaster which creates a natural appearance. Towards the corner, the brick is replaced with glass openings, thus creating maximum openness at the corner of the building. Inspiration for the design of the facade was found in the image of a Venetian palazzo. The stateliness, the composition and ornamentation are translated into a contemporary appearance. The result is a building with a strong architectonic expression, answering the needs and measurements of the location, yet satisfying all the demands that are requested of a modern office building.

Erick van Egeraat

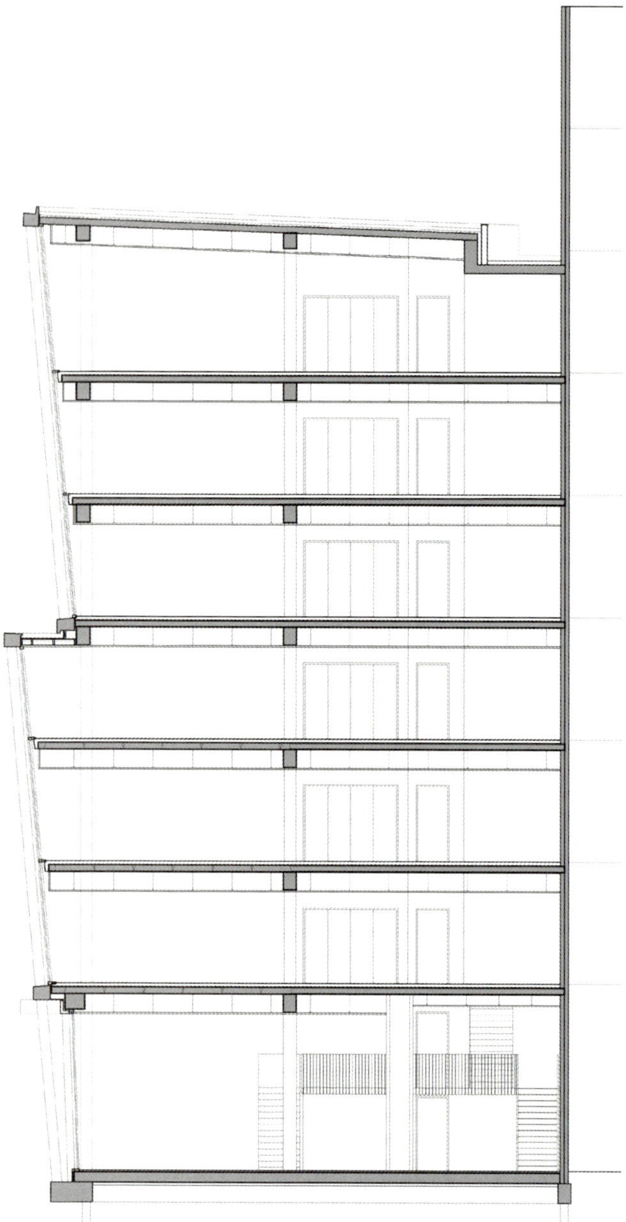

\ section

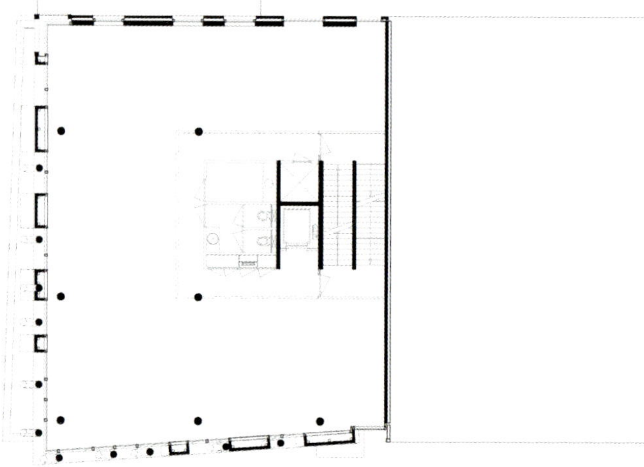

level 5 /

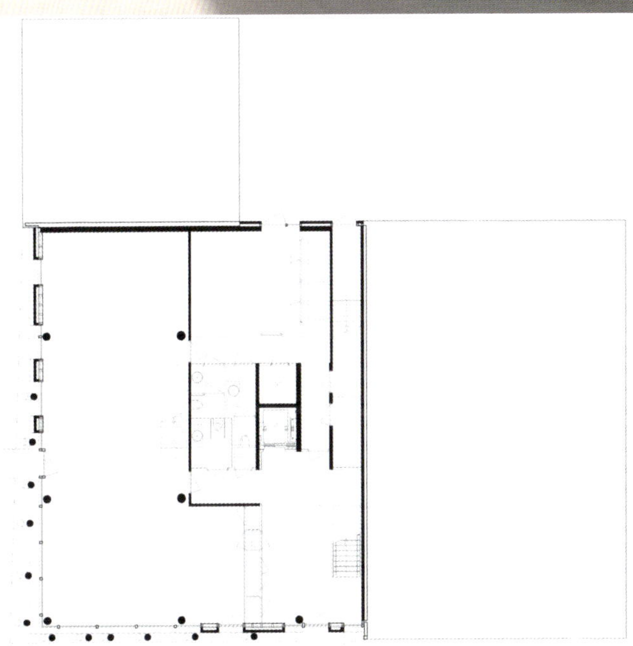

mezzanine floor plan /

Almere
L'stad Haven
Botter
Tjalk
Noordersluis
Schoener
Emmeloord
Enkhuizen
Swifterbant
Schouw-Station
IJssellaan
Zuiderzeelaan
Hypotheek Checkpu

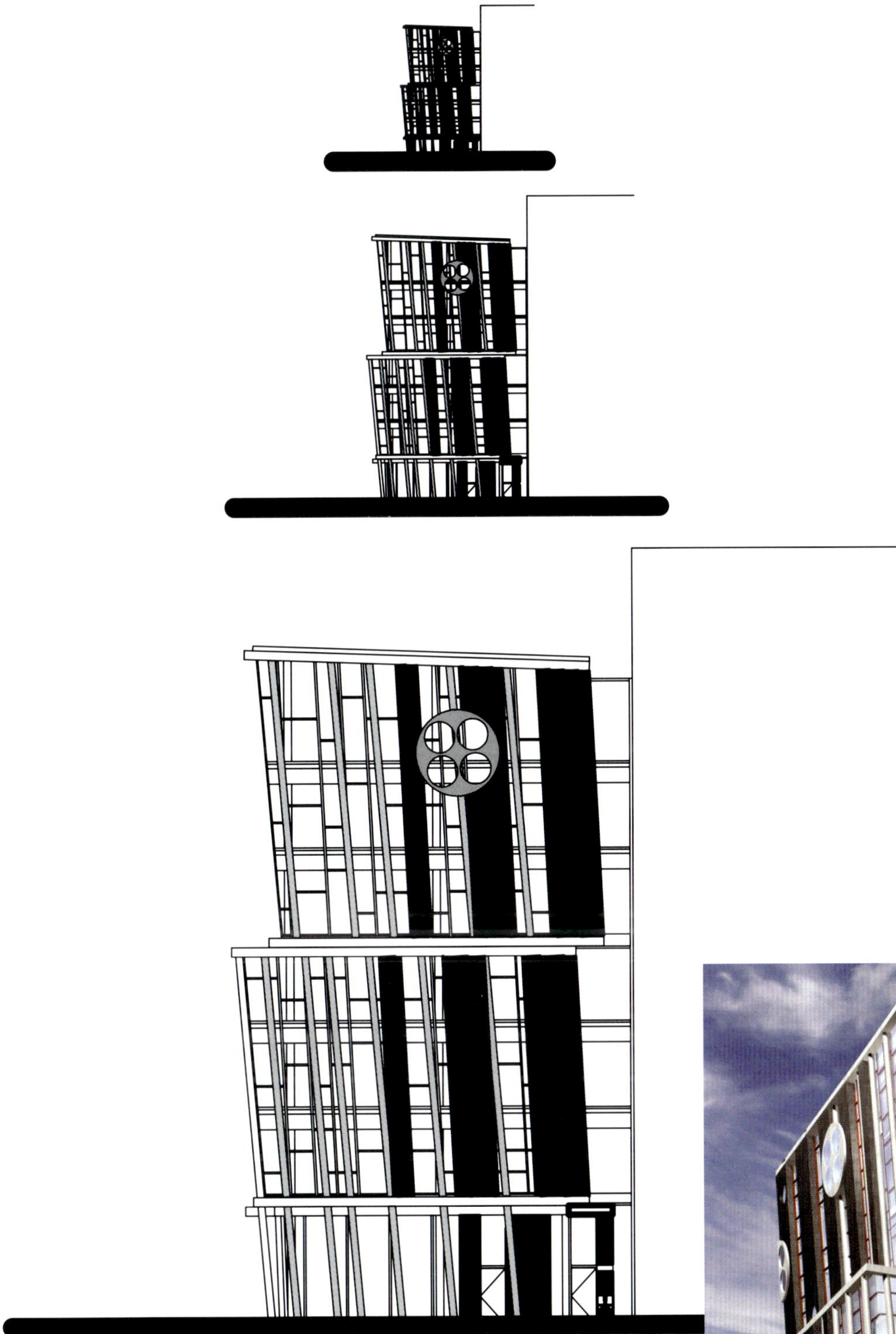

\ facade

final exterior image /

Lelystad, the Netherlands

2006

Rene van Zuuk Architekten

Function_ Office

Site area_ 222m^2

Total floor area_ 6,620m^2

Project architect_ Rene van Zuuk, Kersten Scheller

Structural Engineer_ Van Rossum

Photographer_ Chistian Richters

Client_ Develop-Havelte bv

ZILVER-PARKKADE D

이 프로젝트의 로맨틱한 명칭(부두의 은색 공원)은 렐리스타트(Lelystad)시 질베르파케이드(Zilverparkkade)의 사무실 클러스터라는 위치에서 유래한 것이다. 웨스트 에잇(West 8)은 밀집된 도심을 위해 최근의 경향을 따라 도시계획을 진행했다. 사무소 건물들은 17세기 네덜란드풍의 운하주택에 한가로이 윙크를 보내며 좁은 대지 위에 줄지어 서 있다. 엄격한 구획으로 그토록 촘촘하게 줄지어 있는 상황에서 차별성을 만들어낼 수 있는 요소는 파사드가 유일하다.
4개 파사드 표면은 모두 전체적이든 부분적이든 미리 제조된 콘크리트 구성물들로 덮여있는데, 이것들은 나뭇가지 모양의 구조를 상징적으로 보여준다. 과장된 장식은 무한 연속 패턴을 연구한 결과다.
네덜란드의 그래픽 아티스트 모리츠 에셔(Maurits Escher)의 작품이 그 모델이 되었는데, 한정된 종류의 콘크리트 요소를 적용해 단절 없이 하나로 완결하는 반복적인 패턴을 고안하려면 특별한 기술이 요구된다. 이런 반복적인 요소들이 눈에 잘 띄지 않도록 하는 것이 장인정신이며, 이러한 반복적인 요소들은 직물과 벽지를 위한 순환인쇄패턴의 디자인에서 유추되었다.
파사드의 조직체계는 모서리를 돌며 연속되는데, 사무실 건물이 인접한 건물과 다른 정체성을 가졌다는 것을 인식할 수 있게 한다.
남쪽과 서쪽 입면은 '창문-청소' 프레임을 끼워 넣은 이중 파사드로 되어 있다. 이런 좁은 발코니로부터 얻어지는 부수적인 이점은 화재가 수직적으로 확산되는 것을 막는 기능을 한다는 것이다.
결론적으로 1층 높이의 열저항 기능이 없는 일반 유리창이 적용될 수 있을 것이다. 돌출부는 거대한 햇볕 차단기의 역할을 한다. 나뭇가지 구조물은 밖에서 바라볼 때는 모호한 장면을 연출하지만, 내부의 사용자는 주변을 바라볼 때 거의 방해를 받지 않는다.
북측 및 동측 파사드 위의 콘크리트 구성물들은 단단하게 고정되어 있다. 이런 판넬들 위의 가지 패턴은 고부조 방식으로 새겨지고, 두 가지 다른 혼합재료를 사용함으로써 강조된다. 후면 파사드에는 무늬가 새겨진 콘크리트와 창문으로 이루어진 수평 띠와 같은 교차구조가 있고, 수직적인 틈은 건물을 비대칭적으로 나누고 있다. 이 절단면은 건물기능이 구분되었음을 시각적으로 보여주고, 고정된 콘크리트 판넬의 반복에 미묘한 영향 끼친다.
질베르파케이드 프로젝트는 2004년 착공되어 2005년 말 완공되었다.

The project derives its romantic name(Silver Park Quay) from its location: the office cluster on the Zilverparkkade in Lelystad. West 8 designed the urban plan for the area in conformity with the current trend for condensed city-centres. With an idyllic wink to the 17th century Dutch canal-houses the offices are lined up shoulder to shoulder on narrow plots. In such a compact row with a strict layout the only elements to create a discriminating identity are the facades.

All four facade surfaces in the design are either entirely or partly covered with prefabricated concrete elements, symbolizing a branch-like structure. This blown-up filigree is the result of a study of infinite patterns. The works of the Dutch graphic artist Maurits Escher have been an unmistakable model. Specific skills are required to devise a repetitive pattern that, applied in a limited number of different concrete elements, constitutes a seamless entity. The craftsmanship needed to avoid the repeating units to be too obvious, has an analogy with designing rotation press patterns for textile and wallpaper.

The facade structure continues around the corners, providing the office building a recognizable identity distinct from the adjacent properties.

The south and west elevation have a double layer facade with 'window-cleaners' ledges in between. A profitable side-effect of these narrow balconies is their function as a barrier to prevent vertical spread of fire. As a result storey high panes of ordinary (non-heat-resistant) glass could be applied. The ledges also act as large sun protection slats.

The branch structure provides an obscured vista from the outside, but from the inside the users hardly have any impairment of the view on the surroundings.

The concrete elements on the north and east facade are solid. The branch-pattern on these panels is applied in high relief en is highlighted by using two different shades of aggregates. The rear facade has an alternating structure of horizontal strips of textured concrete and windows, with a vertical cut splitting the building asymmetrically. This incision visualizes the zoning and subtly affects the repetition of the solid concrete panels.

The construction of the Zilverparkkade-project has commenced in 2004 and will be completed by the end of 2005.

Maarten Willems

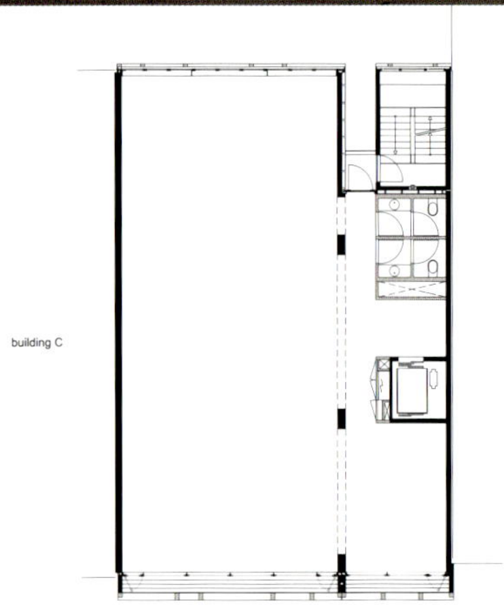

\ 1st - 6th floor plan

\ 7th + 8th floor plan

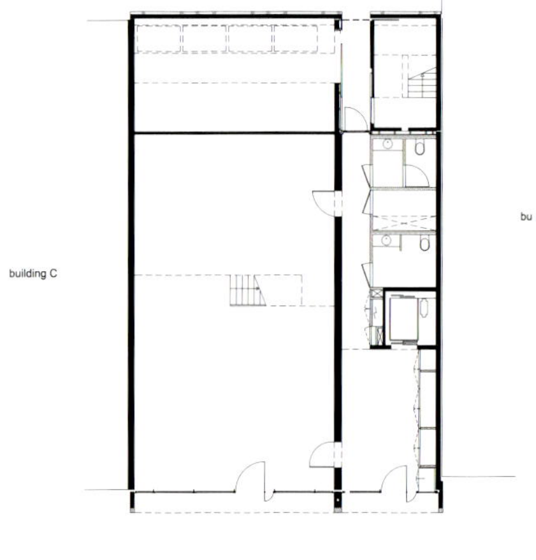

ground floor plan /

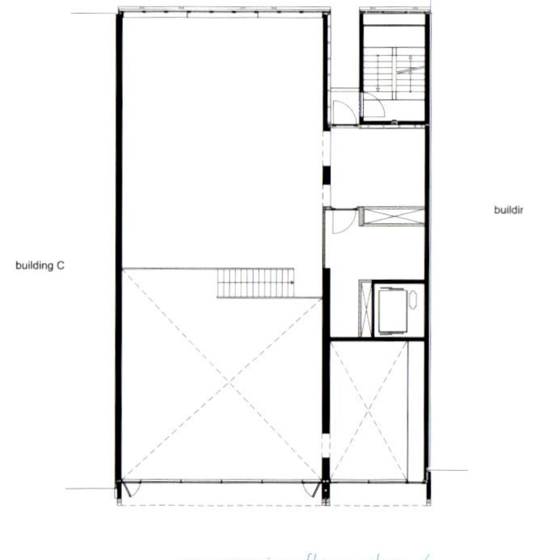

mezzanine floor plan /

Escher, Mauritis Cornelius /

\ model

situation /

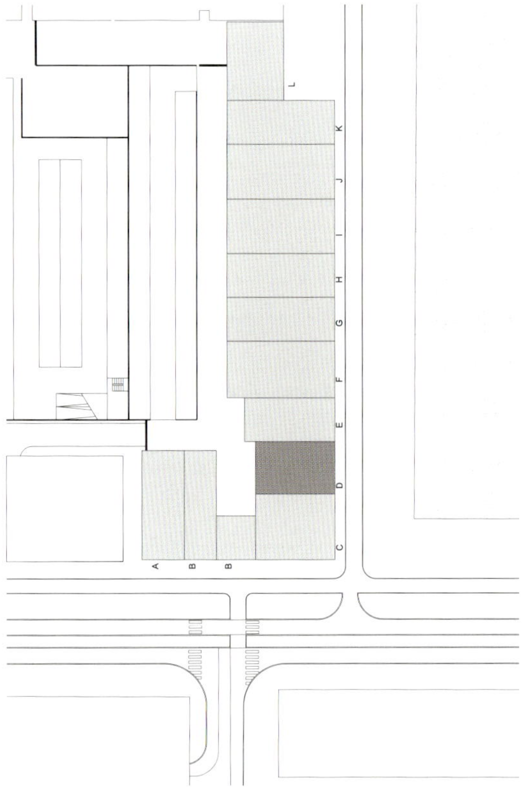

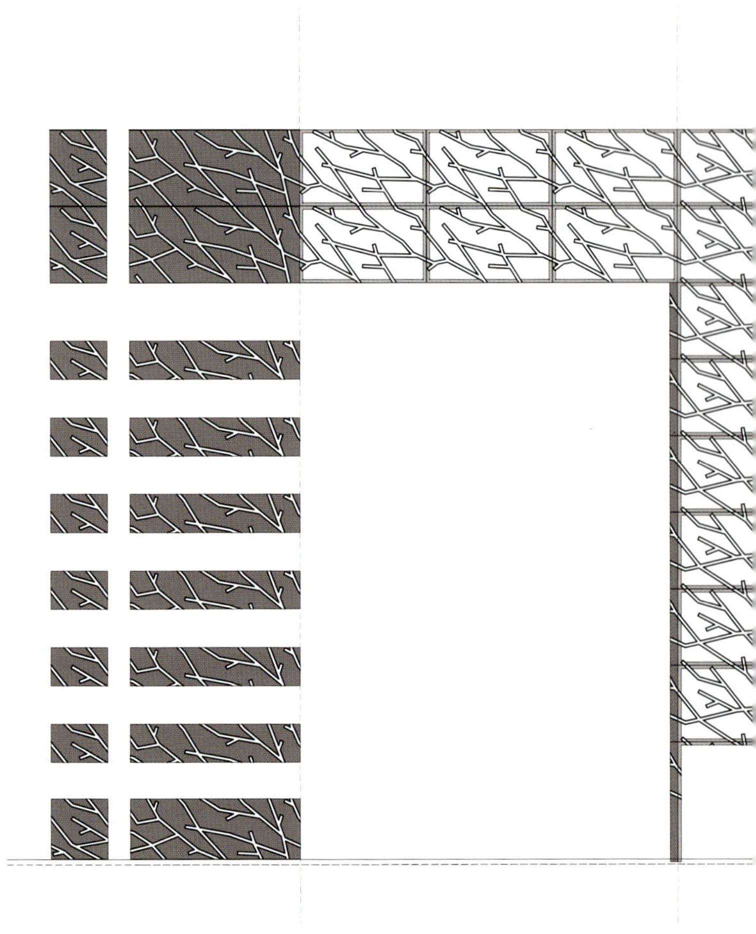

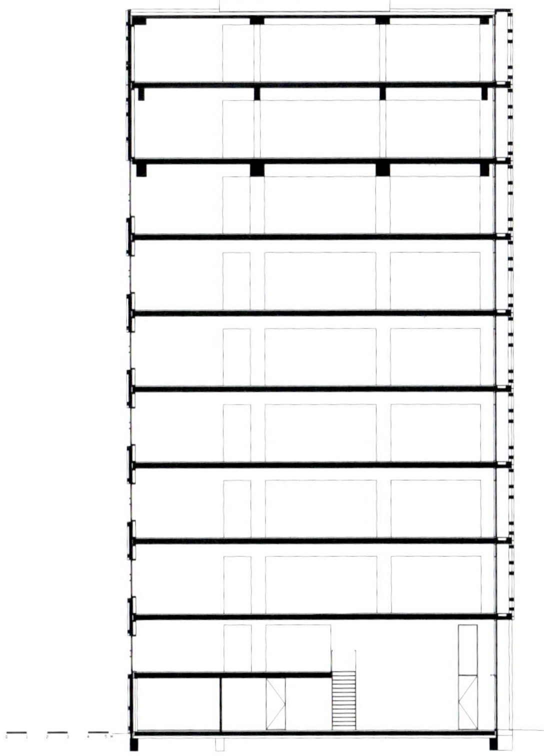

\ section

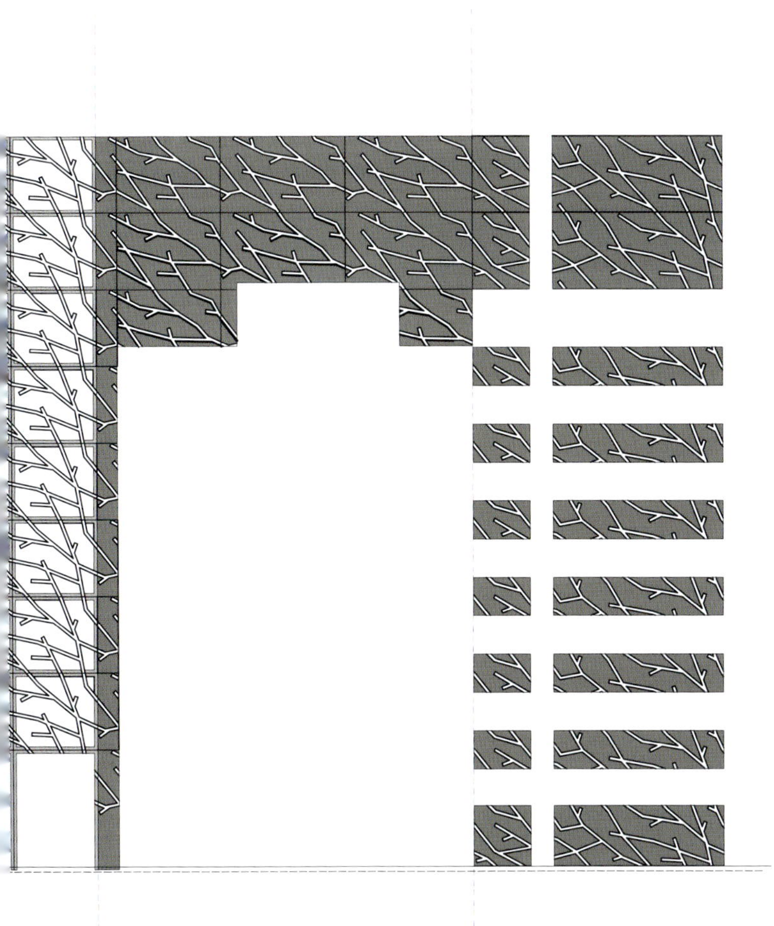

\ pattern

NEIGHBOR OFFICE BUILDINGS IN THE ZILVERPARKKADE

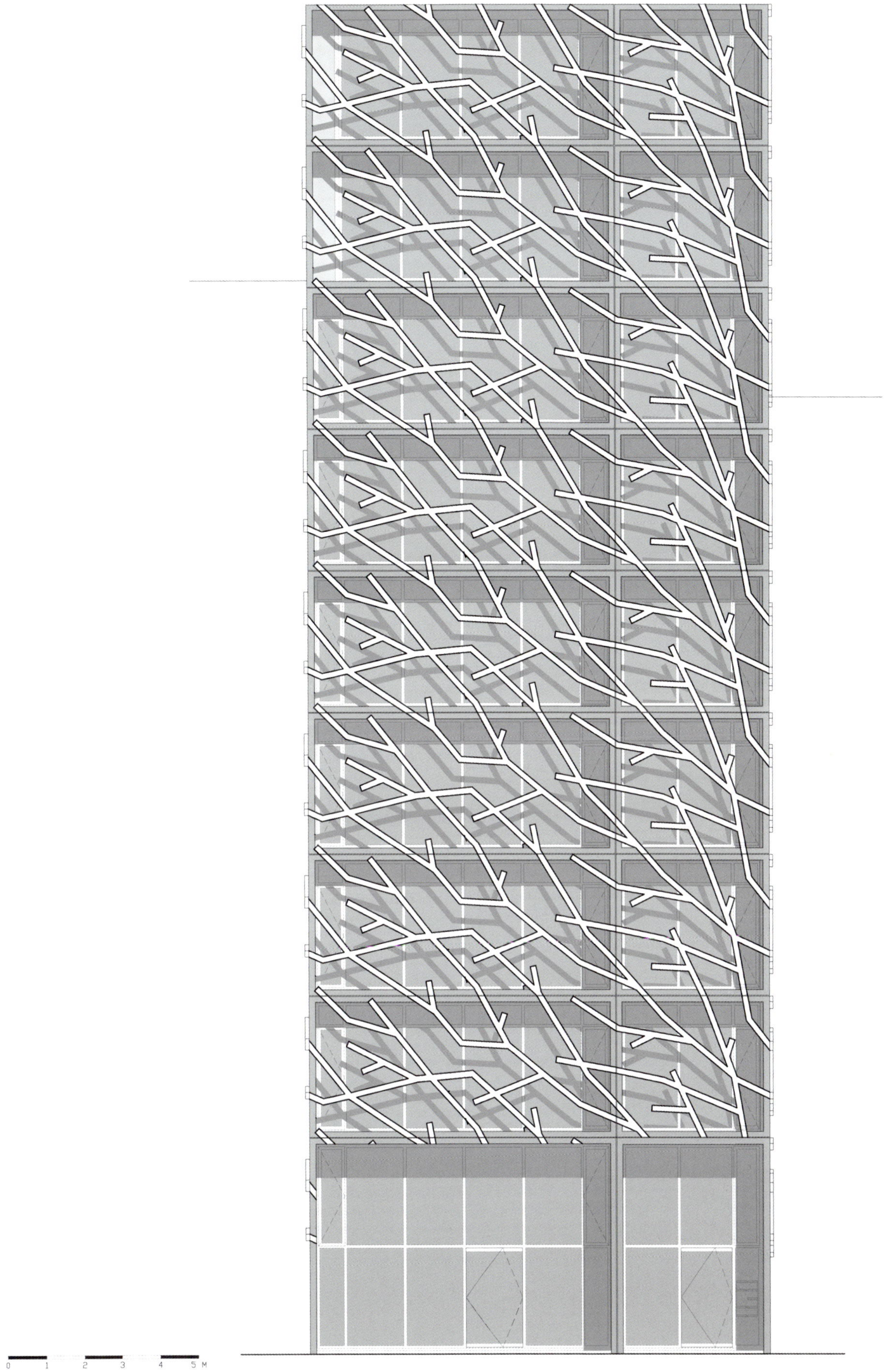

\ south elevation

이중 레이어 파사드

서쪽 파사드는 보존되어야 할 공원을 향한 긴 조망을 제공한다. 커다란 창문은 바닥에서 천장까지 일하는 공간 안에서 보여지는 모든 파노라마를 보여 준다.
유리로 된 파사드를 청소하도록 건축주는 지붕에 청소부를 위한 장치를 설치하는 대신 보조적인 기능을 하는 발코니를 만들고자 하였다. 반면, 도시계획은 건물이 앞으로 내밀거나 뒤로 빠질 수 없도록 건물의 모든 파사드가 같은 면을 갖도록 규정하고 있었다. 해결책은 이중레이어로 구성된 파사드를 만드는 것이었다. 발코니는 이 두 레이어 사이에 있다. 첫 번째 레이어는 떨어지는 것을 방지하고, 물이 스며들 수 있는 콘크리트 나뭇가지 모양의 패턴으로 만들어졌고, 두 번째 레이어는 거대한 유리 스크린으로 구성된다.
북쪽 파사드는 뒤뜰을 마주 보고 있다. 작은 창문은 양면을 모두 청소할 수 있도록 안으로 열리게 된다.

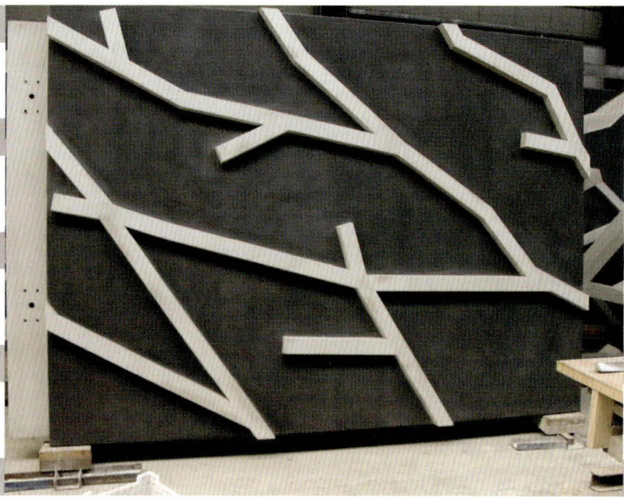

Double layer facade

The South Facade offers long views to a park which should be preserved. Big windows from floor to ceiling bring all this panorama inside the working space.

On the one hand, in order to clean this glazed facade, the client decides to have auxiliary balconies to avoid a window-cleaner device on the roof. On the other hand, the urban plan obligates all the facades of the buildings to be in the same plane, which makes it impossible to go backwards or forwards.

The solution comes with a double layer facade. The balconies are in between these two layers. The first layer is made by a permeable pattern of concrete branches which functions as a falling-protection and the second in a large screen of glass.

The North facade faces a backyard. Smaller windows open towards the inside to clean it on both sides.

Rene van Zuuk Architekten bv.

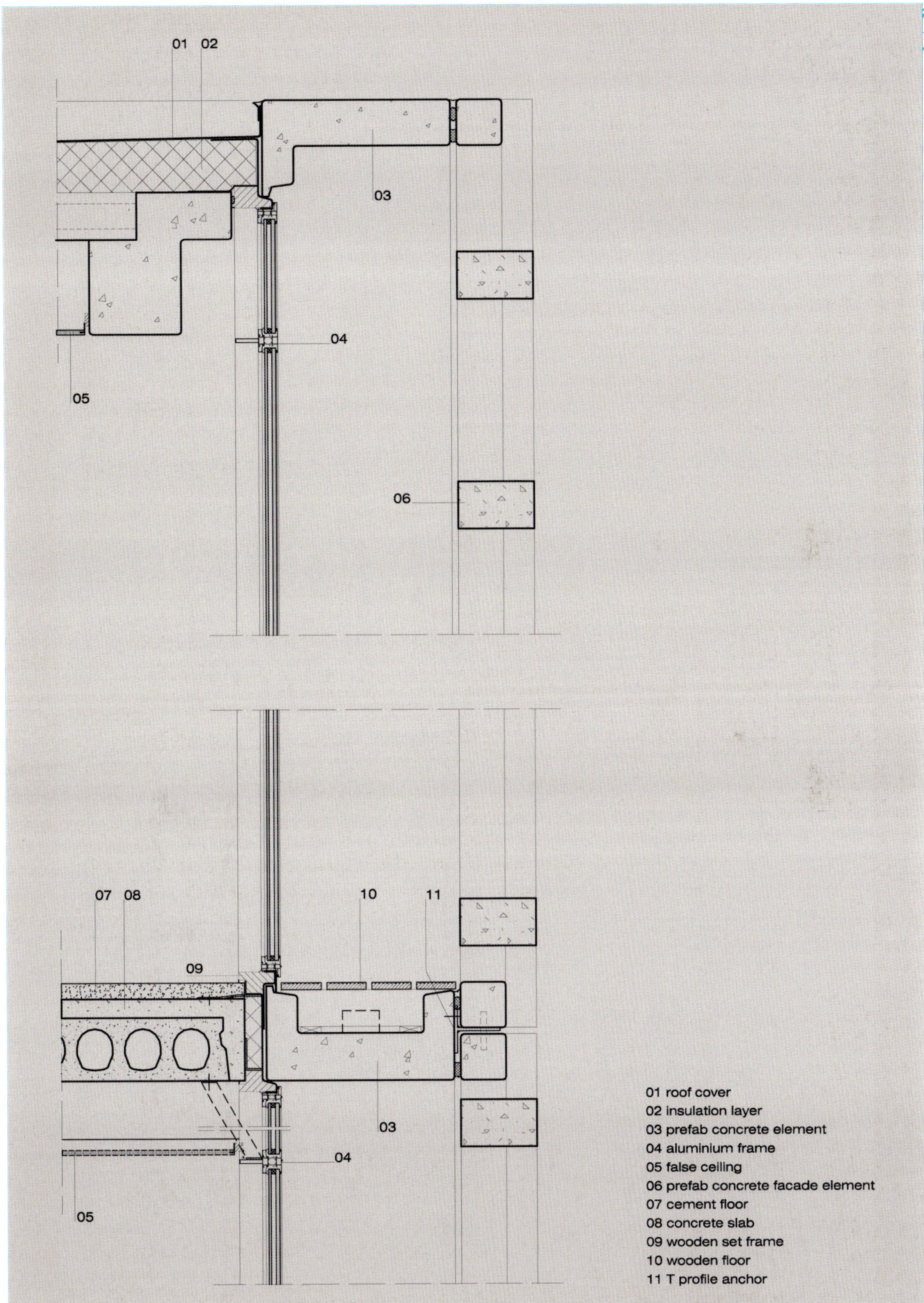

south facade detail

UNIQUE GLASS FACADE IN THE CITY

싱가폴에 지어진 '장난감 미술관(toy museum)'과 프랑스에 지어진 '시트로엥 쇼룸(Citröen showroom)'은 건물이 놓인 지역의 특성이 분명히 다름에도 불구하고, 도심의 주요 상업 거리에 비슷한 층고로 지어졌다는 점과 유리를 이용한 장식적이고 구조적인 파사드를 가지고 있다는 점에서 시각적 유사성을 가지고 있다. 건축이 더 이상 특정 개념이나 유형에 집착하지 않고, 내부의 프로그램과 외부의 환경과 반응하며 만들어내는 디자인의 유연성은 재료에 대한 새로운 구축의 방식을 통해 효과적으로 표현되고 있다. 두 작품에서 건물의 목적을 상징화하여 장식된 유리는 빛의 유입과 밀접한 관계를 가지고 있으며, 실내외 공간 사이의 경계를 형성하는 파사드는 일종의 실내외환경 조절 시스템으로서 역할을 담당하는 동시에 형태를 구축하는 수단으로 사용되고 있다.

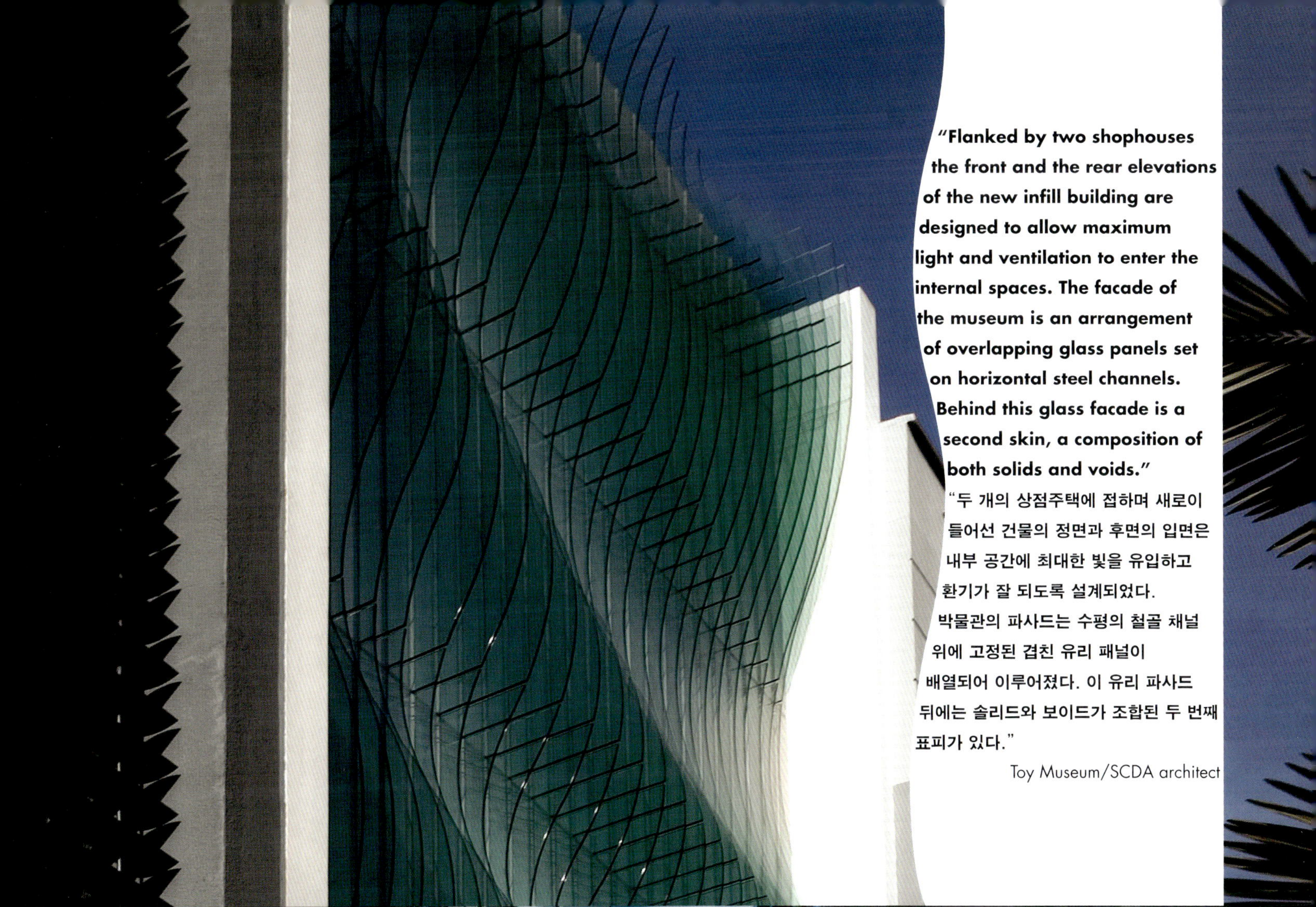

"Flanked by two shophouses the front and the rear elevations of the new infill building are designed to allow maximum light and ventilation to enter the internal spaces. The facade of the museum is an arrangement of overlapping glass panels set on horizontal steel channels. Behind this glass facade is a second skin, a composition of both solids and voids."

"두 개의 상점주택에 접하며 새로이 들어선 건물의 정면과 후면의 입면은 내부 공간에 최대한 빛을 유입하고 환기가 잘 되도록 설계되었다. 박물관의 파사드는 수평의 철골 채널 위에 고정된 겹친 유리 패널이 배열되어 이루어졌다. 이 유리 파사드 뒤에는 솔리드와 보이드가 조합된 두 번째 표피가 있다."

Toy Museum/SCDA architect

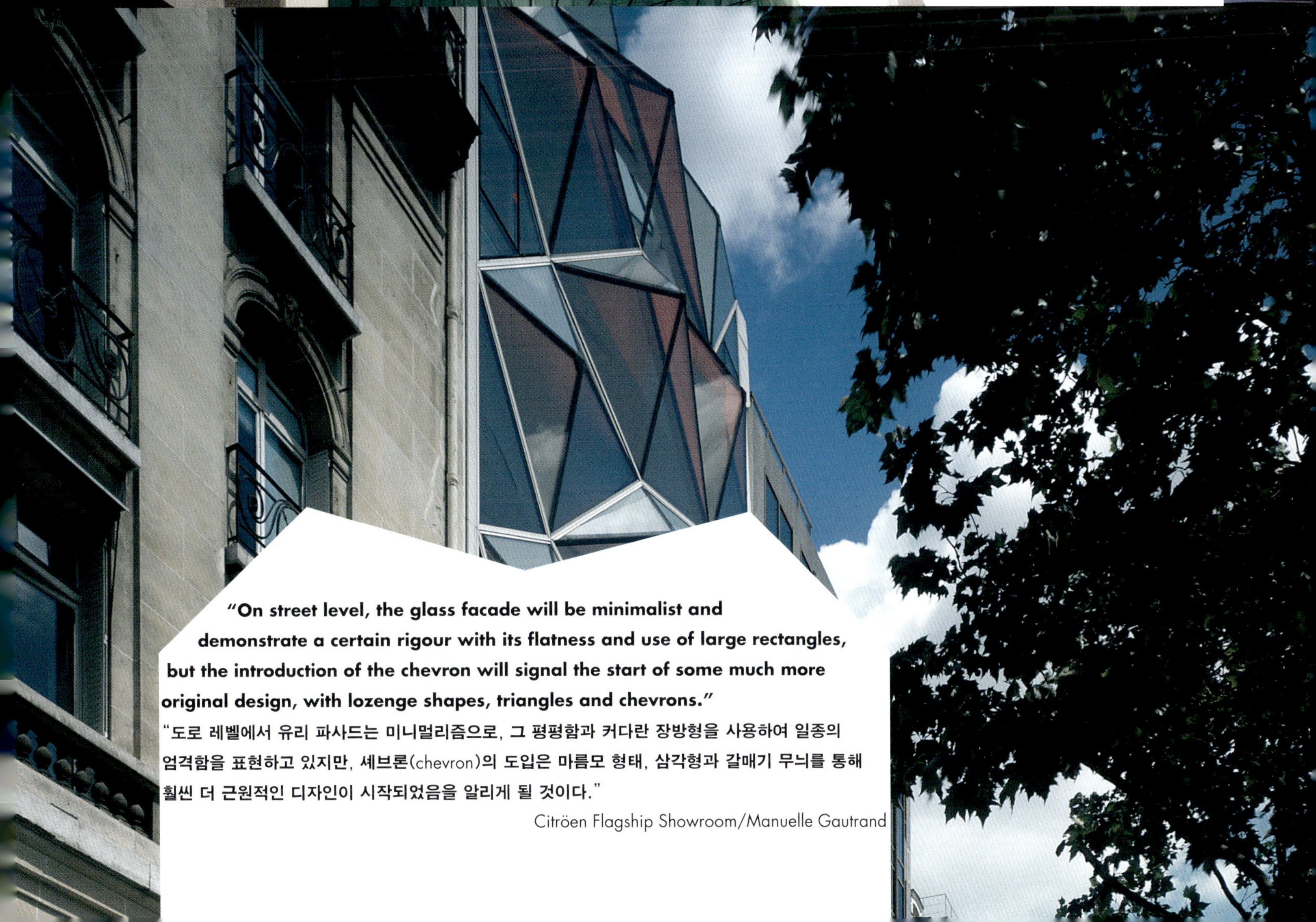

"On street level, the glass facade will be minimalist and demonstrate a certain rigour with its flatness and use of large rectangles, but the introduction of the chevron will signal the start of some much more original design, with lozenge shapes, triangles and chevrons."

"도로 레벨에서 유리 파사드는 미니멀리즘으로, 그 평평함과 커다란 장방형을 사용하여 일종의 엄격함을 표현하고 있지만, 셰브론(chevron)의 도입은 마름모 형태, 삼각형과 갈매기 무늬를 통해 훨씬 더 근원적인 디자인이 시작되었음을 알리게 될 것이다."

Citröen Flagship Showroom/Manuelle Gautrand

Singapore

2002-2005

SCDA architects

Function_ Museum

Site area_ 155.3m^2

Total floor area_ 652m^2

Project Architect_ Chan Soo Khian, Edward Lau

Structural Engineer_ Arup

Photographer_ Albert Lim

TOY MUSEUM

싱가포르의 장난감 박물관은 유서깊은 래플즈(Raffles) 호텔의 측면 출입구를 바라보는 시아(Seah) 거리에 위치한다. 예전에 상점주택이 있었던 대지의 폭은 5.5미터이고, 그 길이는 27.5미터이다. 신축 건물의 높이는 계획 규정에 따라 5층으로 제한되어 있었다.
두 개의 상점주택에 접하며 새로이 들어선 건물의 정면과 후면의 입면은 내부 공간에 최대한 빛을 유입하고 환기가 잘 될 수 있도록 설계되었다. 박물관의 파사드는 수평의 철골 채널 위에 고정된 겹친 유리 패널이 배열되어 이루어졌다. 이 유리 파사드 뒤에는 솔리드와 보이드가 조합된 두 번째 표피가 있다. 보행자 레벨의 두 층 사이에는 선큰정원 위로 난 보이드를 가로지르는 차폐된 유리 통로가 있다.

주출입구 로비는 전시공간으로 2층 높이로 만들어져 아래의 카페를 내다보게 된다.
내부에는 무주공간이 생겨났고 모든 슬래브와 벽체는 원래의 노출콘크리트로 되어 있다. 설비는 서로 맞물린 공간을 연결하는 유일한 수직 요소인 '서비스 코어' 내부에 들어 있다. 단면상 번갈아 형성된 층은 공간적으로 다양해서, 거리를 내려다볼 수 있는, 빛으로 충만한 대규모 공간 내부에 나란히 놓인 작은 공간들이 인테리어가 조화를 이룬 '장난감 상자'가 되게 한다. 옥상은 도시 경관을 내다보는 조망을 지닌 개방된 테라스로 설계되었다. 거리 위로 높게 세워진 장난감 박물관은 각종 행사나 특별한 기능을 위해 사용될 것이다.

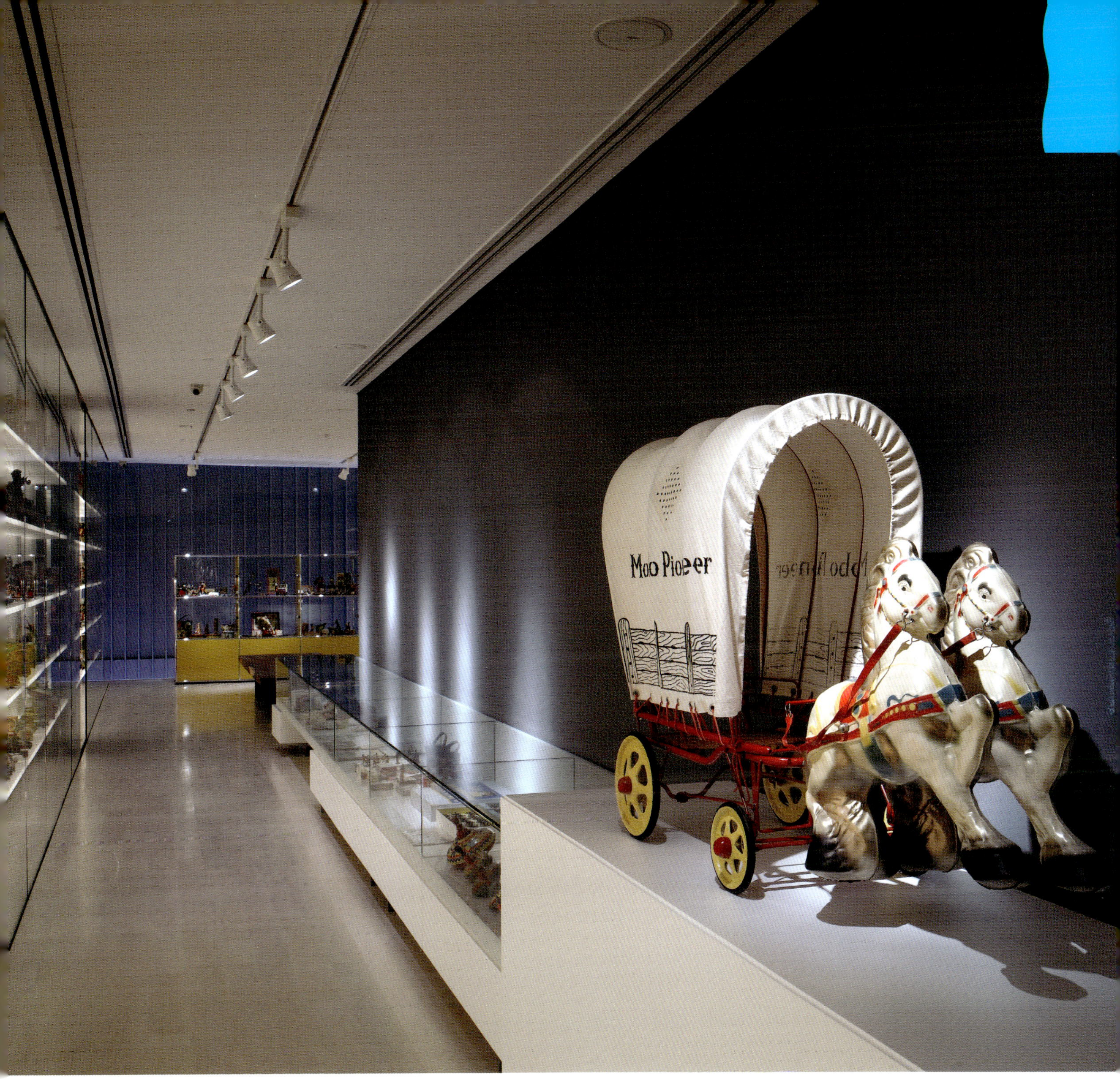

Singapore's toy museum is located in Seah Street, facing the side entrance to the historic Raffles Hotel. The site, a former shophouse, is just 5.5m wide and 27.5m in length. The height of the new building was constrained by planning regulations and it was limited to five storeys. Flanked by two shophouses the front and the rear elevations of the new infill building are designed to allow maximum light and ventilation to enter the internal spaces. The facade of the museum is an arrangement of overlapping glass panels set on horizontal steel channels. Behind this glass facade is a second skin, a composition of both solids and voids. In between the two layers, at the pedestrian level, is a covered glass walkway, which traverses the void above a sunken courtyard. The entrance lobby doubles as an exhibition space, and overlooks the cafe below.

A column free space is created internally and all floors and walls are of in-situ fairfaced concrete. Services are contained within a 'service core', which is the only vertical element connecting the interlocking spaces.

In section alternate floors vary spatially, permitting smaller spaces to be 'toy boxes' with controlled interiors that are juxtaposed within large-scale volumes of light filled space overlooking the street. The rooftop is designed as an open terrace with views over the urban landscape. High above the street, it will be used for events and functions.

SCDA architects

구조

SCDA 설계사무소는 싱가포르 시아(Seah) 거리에 위치한 장난감 미술관을 위해 강렬하면서도 가물거리며 흘러가는 듯한 파사드를 만들어내기 위한 독창적인 아이디어를 생각해 냈다. 전통적인 상점주택 사이의 좁은 대지라는 위치에 대해, 건축가는 박물관에 물결치는 형태의 유리 파사드를 만들려고 하였다. 값비싼 곡면유리를 사용한 형태를 연구한 후에 일련의 유리 지느러미 형상을 지닌 안이 새로운 개념으로 자리 잡았고, 시선을 끄는 형태를 만들어냈다. ARUP이 이러한 개념을 실현하기 위해 참여했다. 유리 지느러미(the glass fins)를 구조재로 사용하는 방법을 기초로 한 해결책을 발전시켰다. 이것을 통해 구조 부재를 사용하지 않을 수 있었고, 알루미늄 삽입재의 두께를 최소화했다. ARUP은 유리 지느러미를 구조재로 활용할 수 있는 방안을 개발했다. 형태를 만들기 위해 각 지느러미는 컴퓨터로 제어되는 유리 절단기를 사용해 통 유리판에서 절단되었고, 유리판들은 강화, 열 처리되었다. 합판 유리 및 서냉 유리를 포함해 다른 유리의 조합이 고려되었지만 건축주는 갈라짐 효과의 부정적인 영향을 받아들일 수 없다고 생각했고, 서냉 유리의 사용으로 인한 열 파괴가 일어날 위험성이 너무 높았다. 곡면을 강조하면서 유리 지느러미의 물결치는 테두리를 비추는 조명 시스템을 추가했다. 디자인 상에 안전함을 결합하면서 유리 바닥이 파사드를 지지한다. 또한, 화재점검패널은 비상시에 파사드가 개방되도록 한다. 이 패널은 지느러미의 선형을 차단하지 않도록 조정되었다.

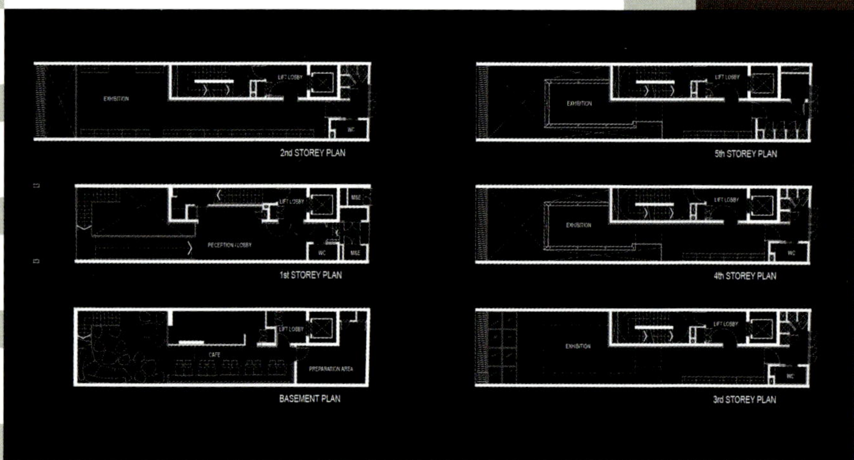

Structural design |

SCDA architects struck upon an ingenious concept to create a vivid, shimmering and flowing facade for the Toy Museum in Seah Street, Singapore. Located on a narrow plot amongst traditional shop houses, the architect's intent was to form a glass facade with a rippling form. After exploring forms using expensive curved glass, a scheme based on a series of shaped glass fins was conceptualised, resulting in an eye-catching form. Arup was engaged to realise this concept. A solution was developed based on using the glass fins as structural members. This avoided the need for framing members and minimised the thickness of the aluminium infills.

Arup developed a solution whereby the glass fins were used as structural members. To create the form, each fin was cut from stock glass sheet using computer controlled glass cutters. The sheets were then toughened and heat soaked. Other glass configurations were considered, including laminates and annealed glass, but the client considered the negative visual impact of delamination to be unacceptable. Furthermore, the risk of thermal breakage resulting from the use of annealed glass was too high. Arup also added a lighting system to illuminate the waved edges of the glass fins, emphasising their curvature. Incorporating safety into the design, a glass floor supports the facade. Furthermore, a fire access panel allows the facade to be opened in the case of an emergency. Arup worked to ensure that the panel did not disrupt the line of the fins.

Arup

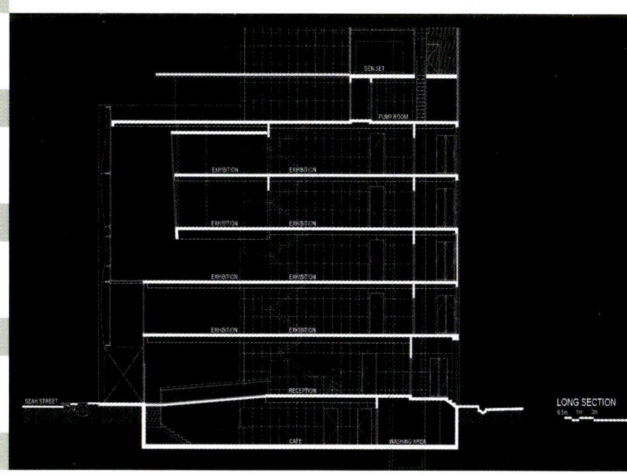

typical glass fin profile section

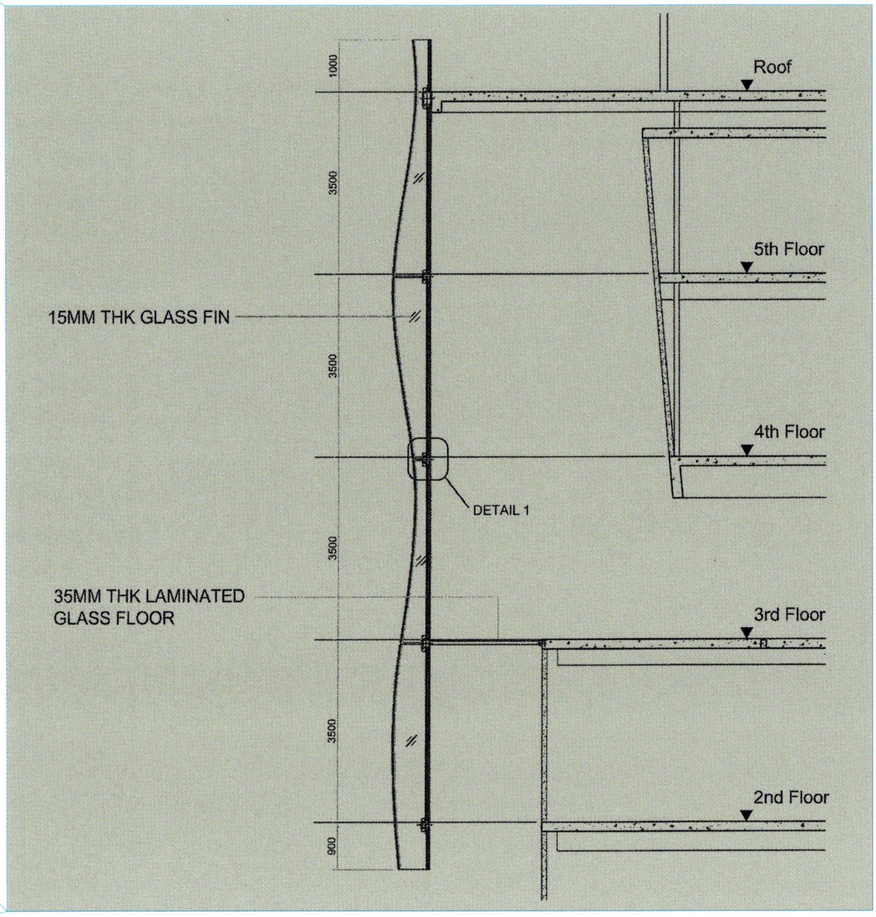

connection detail typical beam vertical section

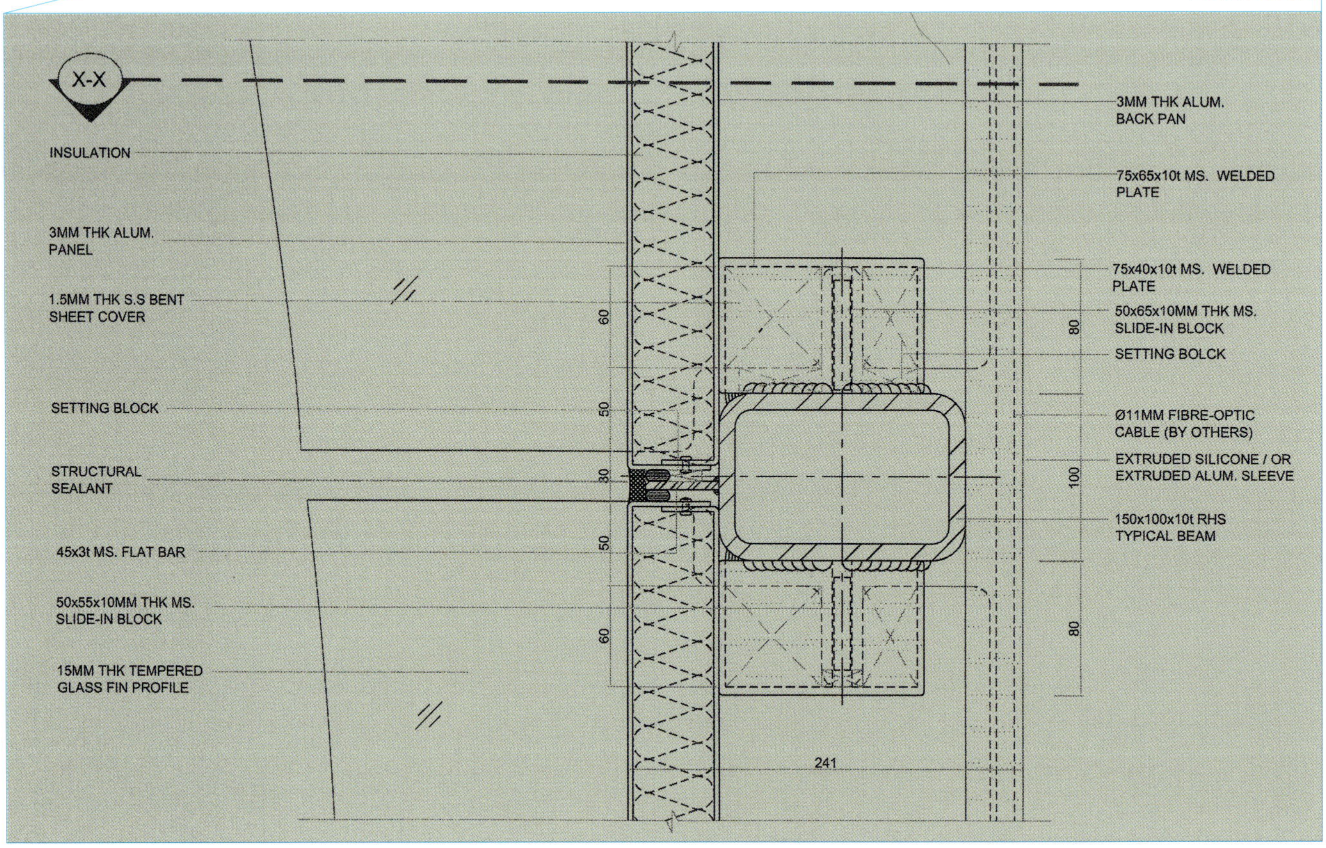

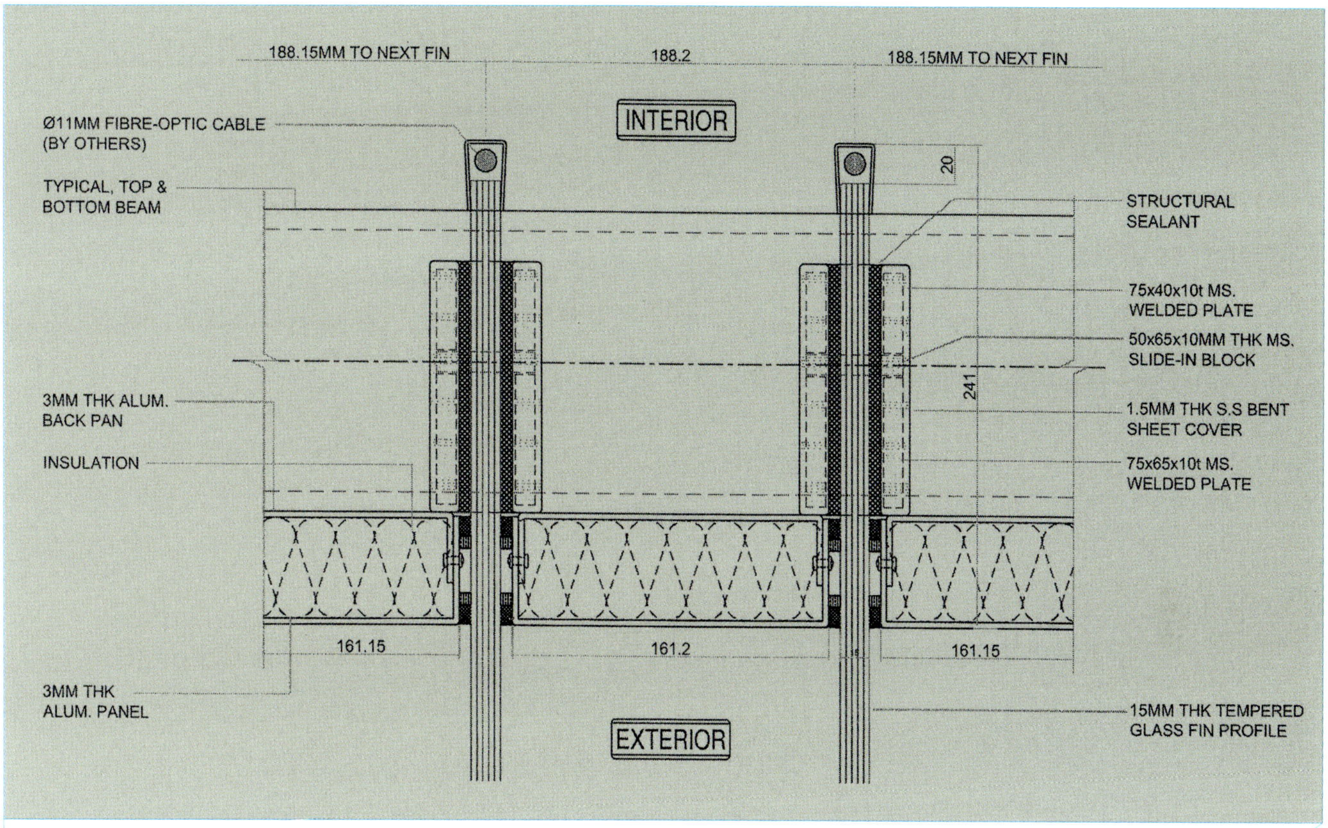

connection detail between beams
horizontal section

Paris, France

2002-2007

Manuelle Gautrand

Function_ Showroom
Total floor area_ 1,200m^2
Project architect_ Manuelle Gautrand, Anne Feldmann
Structural Engineer_ Khephren
Engineers_ Arnaud de Bussière(envelope), Alto(fluid), Lucigny-Talhouet(economist), Lamoureux(acoustic), Labeyrie(multimedia), Casso(security), Spectat(mobil equipment)
Photographer_ Philippe Ruault, Manuelle Gautrand office
Credit_ New Tone(Model), Platform(3D images)
Client_ Automobiles Citröen

CITRÖEN FLAGSHIP SHOWROOM

새로운 시트로엥(Citröen) 쇼룸은 샹제리제(Champs Elysées) 거리 42호에 자리 잡게 될 것이다. 시트로엥은 앙드레 시트로엥(Andre Citröen)이 1920년대에 상점을 시작할 때부터 그 땅을 소유하고 있었다. 기존의 그의 쇼룸은 아름답고, 인테리어는 매우 극적이었으며, 유리로 된 장방형의 파사드는 아름다운 비례로 구성되어, 아주 미니멀적이고 현대적이었다.
도로 레벨에서 유리 파사드는 미니멀리즘으로, 그 평평함과 커다란 장방형을 사용하여 일종의 엄격함을 표현하고 있지만, 셰브론(chevron)의 도입은 마름모 형태, 삼각형과 갈매기 무늬를 통해 훨씬 더 근원적인 디자인이 시작되었음을 알리게 될 것이다. 건물을 더 높이 올려다볼수록 디자인에 새로운 깊이를 더하는 프리즘의 도입으로 건물은 더욱 입체적으로 변화한다. 마침내 새로운 건물의 가장 위층은 그 복잡함 속에 오리가미(origami)를 연상케 하는 거대한 유리 조각처럼 변한다. 갈매기 무늬는 조심스럽게 남아 있지만, 경계는 불명확해지고, 전체 형상 속에서 드러나는 잠재적인 것으로 변한다. 흥분해 마지않을 이번 프로젝트는 건축물과 미술조각품의 중간 형태로 자리 잡는다.
우리는 원래 브랜드의 상징적인 색인 붉은색을 유리패널에 사용하려고 마음먹었지만, 그렇게 하면 외부에서 볼 때 너무 요란한 것이 될 것이라는 결론을 내렸다. 건물이 샹제리제의 주변 건물들과 어울리지 못할 것이라는 약간의 우려가 있어서, 우리는 필터를 만들고 붉은색이 외부로 드러나지 않도록 숨겼다. 이것은 유리마감재 내부에 정교하게 시공된, 완전히 독창적인 필터로, 태양열이 통과되는 것을 최소화하고, 건물 내부에 영롱한 백색 진주 빛을 만들어 내게 된다. 붉은색은 브랜드의 상징적인 색을 반영하면서 건물의 내부에서 보여지게 된다.
건물의 주요 기능은 자동차를 전시하는 것이고, 우리는 이 중요한 목적을 공간의 형태 자체로 표현하기 원했다. 건물의 형태는 자동차의 형상에서 영감을 얻은 것인데, 단순히 전면, 지붕, 후면을 가진 물체가 아니라 자동차의 형태와 유사한 요소들을 연결하는 곡선과 유동성을 통해 모듈화 한 것으로서, 장소와 제품 사이의 통일성을 만들며 실내 환경을 풍요롭고 다채롭게 만든다.
자동차들을 전시할 수 있도록 만들어진 자동차가 하나씩 놓인 8개의 원형 플랫폼은 중앙의 마스트에 연결되었다. 플랫폼의 지름은 6미터이며, 각 플랫폼은 자동차의 전면을 볼 수 있게 되어 있고, 거울로 처리된 베이스는 하부의 자동차를 비춰준다. 방문객들은 전시장 주변을 돌며 자동차를 휘감으며 이어지는 계단과 통로로 이끌리게 될 것이다. 우리는 사람들이 와서 시간을 보낼 수 있도록 일종의 미술관이나 문화시설처럼 만들기 위해 노력했다. 건물에는 사람들을 건물의 꼭대기로 올려줄 전망 엘리베이터가 설치될 것이고, 파리와 파리의 하늘이 선사하는 특별한 전망을 즐기게 될 것이다.

The new Citröen showroom will be at number 42 Champs Elysées; Citröen have owned the site since Andre Citröen set up shop there in the 1920s. His original showroom was beautiful, the interior was extremely theatrical, and the glass rectangle facade beautifully proportioned, very minimalist and contemporary.

On street level, the glass facade will be minimalist and demonstrate a certain rigour with its flatness and use of large rectangles, but the introduction of the chevron will signal the start of some much more original design, with lozenge shapes, triangles and chevrons. The higher up the building one looks, the more three-dimensional it will become with the introduction of prisms that will bring new depths to the design. Finally, the top section of the new building will be like a great glass sculpture, recalling origami in its complexity. The chevron will remain present yet discreet, becoming less defined and more suggested in the overall form, and almost subliminal, in this exciting project, midway between a building and a fine art sculpture.

We originally conceived the use of red, the brand's signature colour, in the glass panels but we decided it would be too bright from the outside. There were some concerns about the building not harmonising with its neighbours on the Champs Elysées, so we've created a filter that on first sight, masks the red colour from the exterior. This totally original filter, which is cleverly constructed inside the finished glass, also minimises the heat of the sun passing through, and will also create a diaphanous pearly white atmosphere inside the building. The red colour can still be seen from the inside of the building, reflecting the brand's signature colours.

The main role of the building is as a place to show cars, and we wanted to express this primary aim in the form of the space itself. The shape of the building itself is inspired by the shape of a car, it's not an object with a front, a roof and a rear, but something moulded with curves and fluidity, that links the front, roof and rear with a continuity that is like the form of a car itself, creating unity between the place and the product, and makes a rich and complex interior.

To display the cars themselves, attached to a central mast will be eight circular platforms each of which will take a car. The platforms will be six metres in diameter, and each one turns to show off the car on all sides and has a mirrored base to reflect the car below. Around the display, the public will be led by a series of staircases and walkways that spiral past the cars. We were trying to create something like a museum or a cultural building, a space which would encourage people to spend time there. There will be a panoramic lift to take people to the top of the building, and they will be able to enjoy an exceptional view of both Paris and the sky.

Manuelle Gautrand

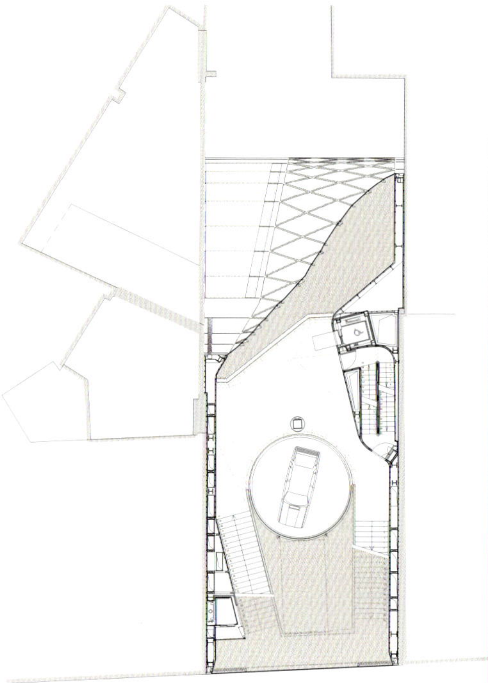

\ 1st floor plan

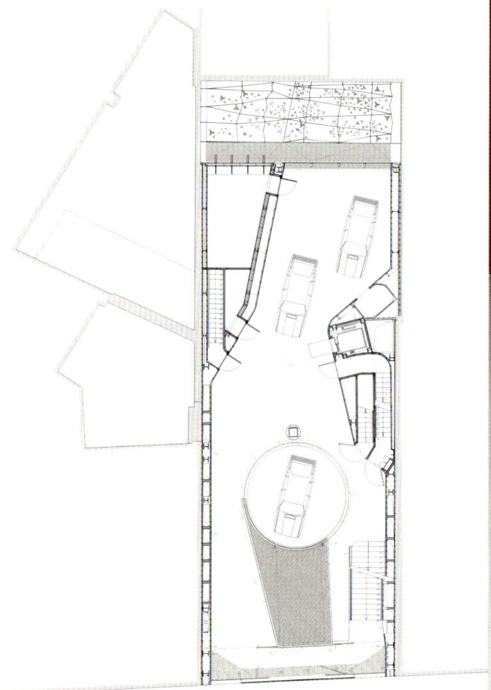

\ ground floor plan

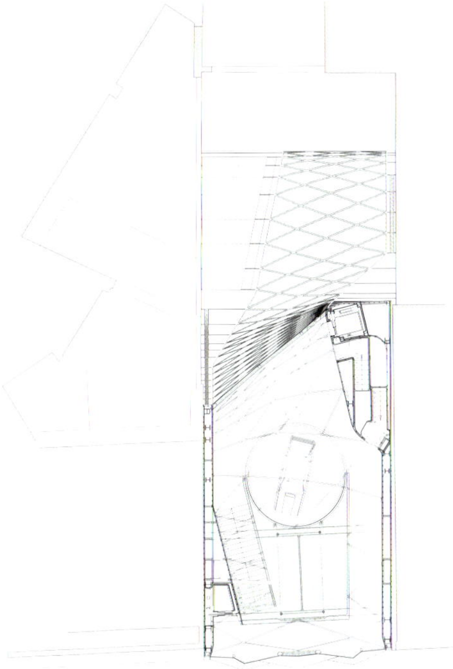

3rd floor plan /

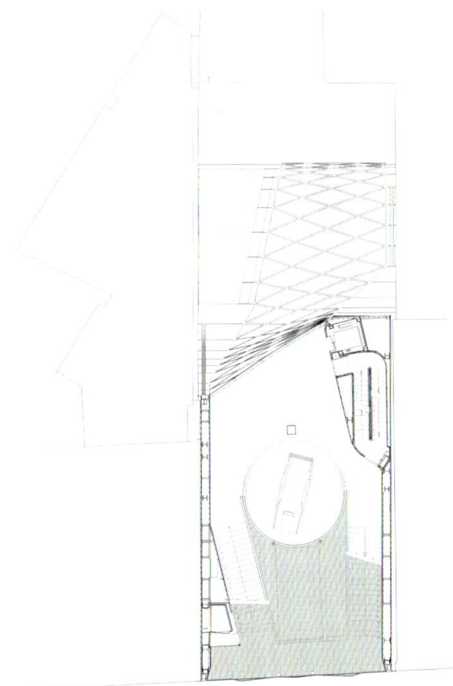

2nd floor plan /

578 DXX 78

\ longitudinal section

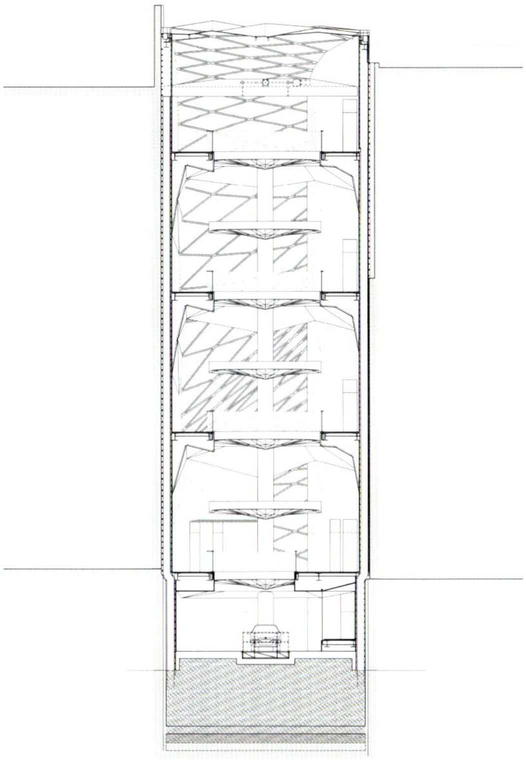

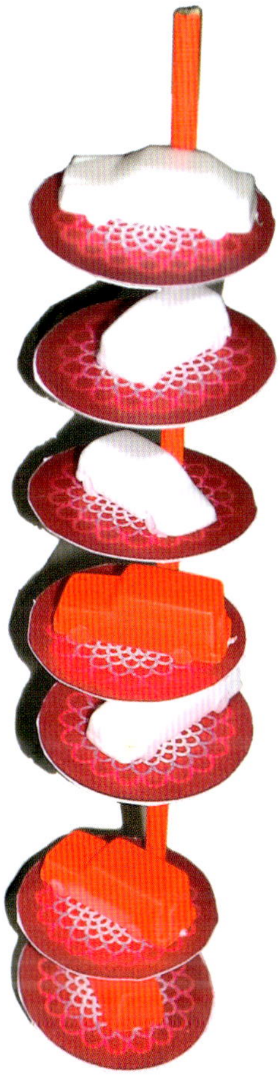

model /

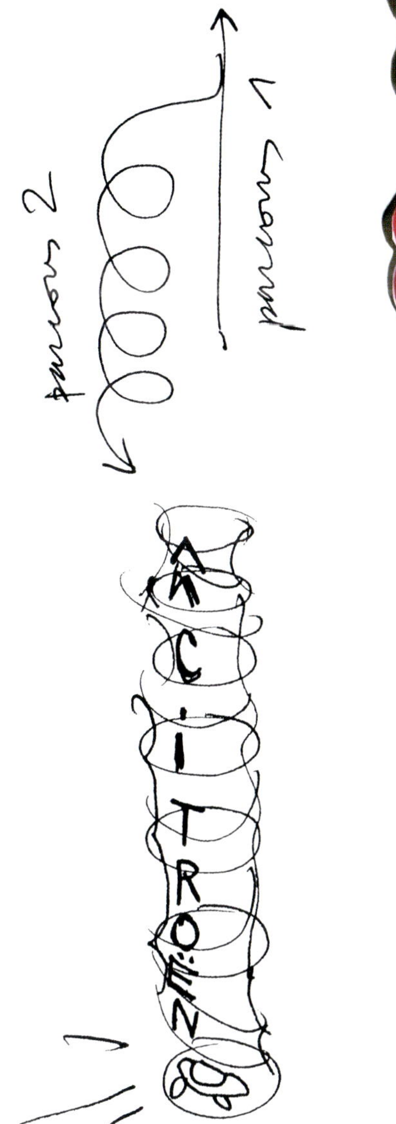

\ sketch

\ situation

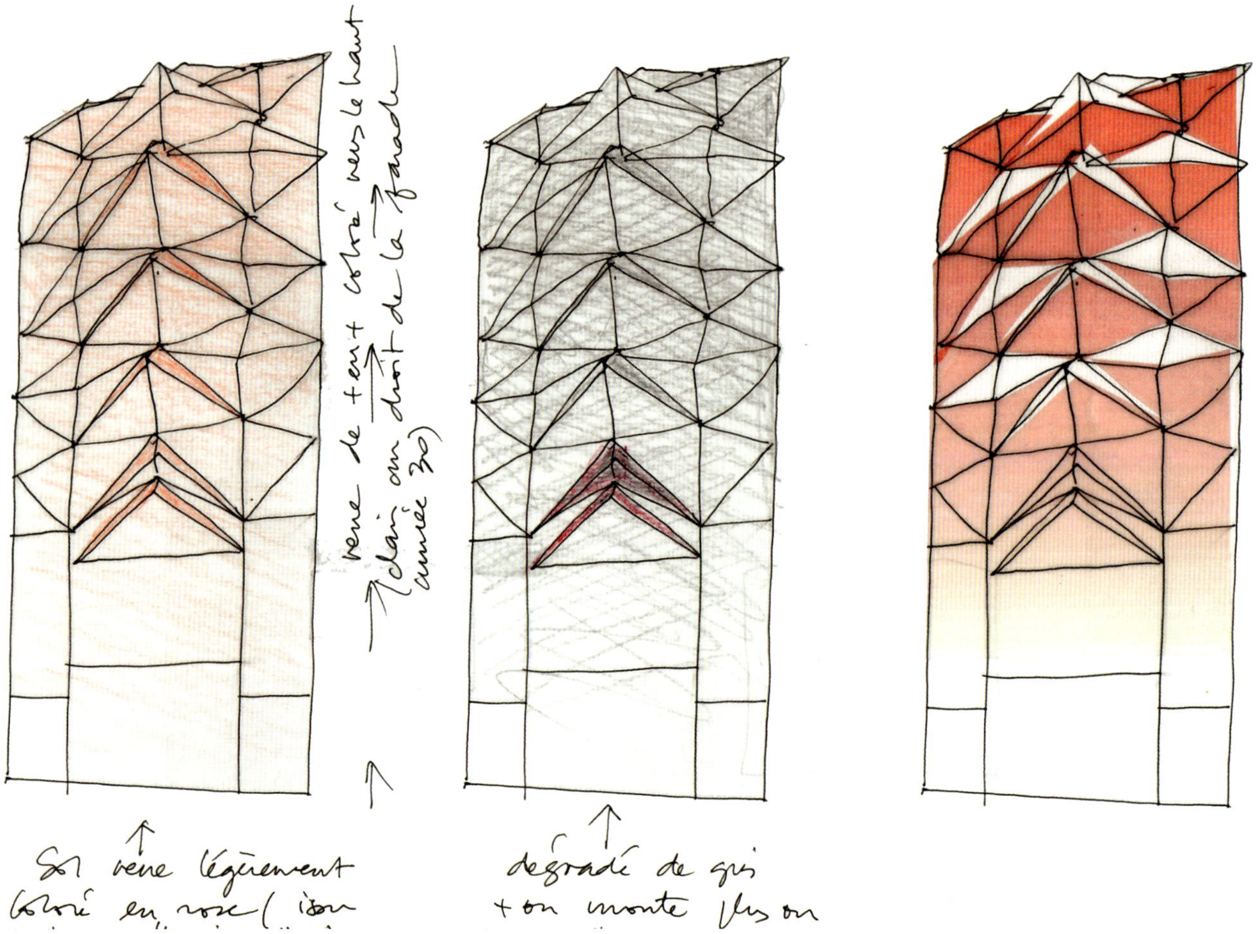

\ sketch

ZARA
±27°C

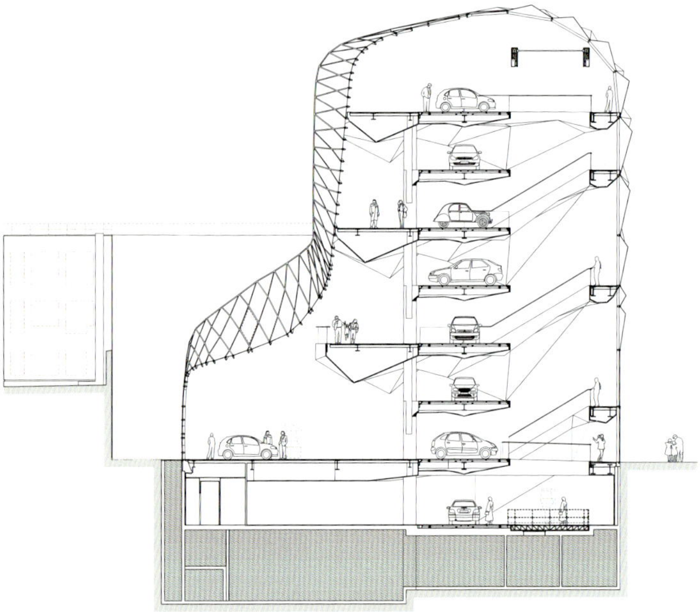

\ cross section

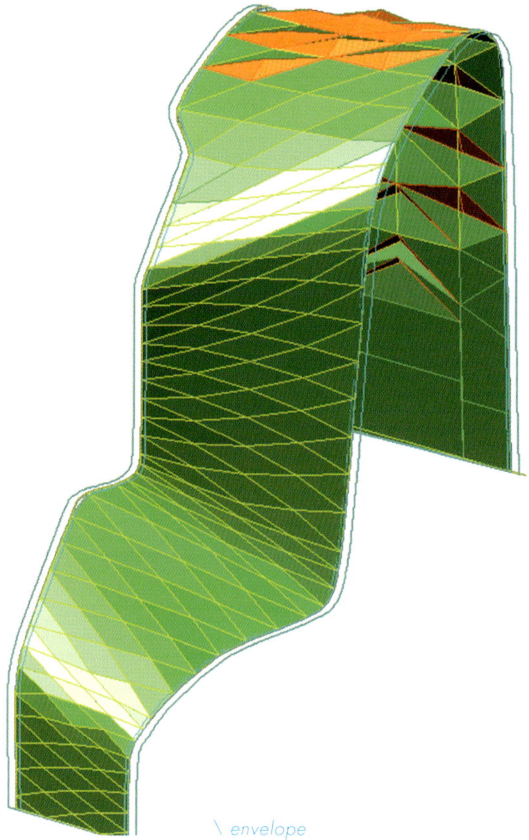

\ envelope

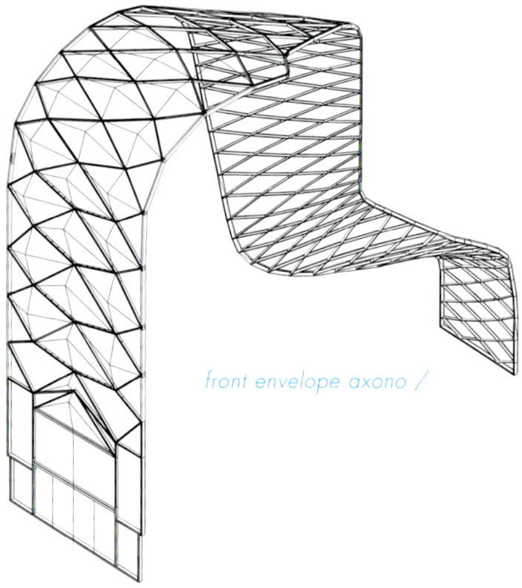

front envelope axono /

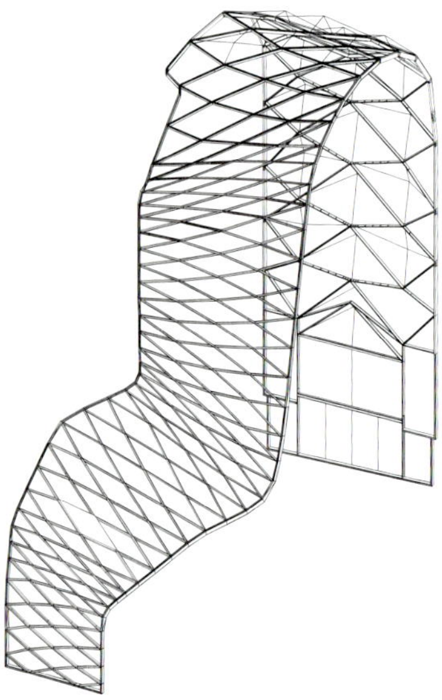

\ back envelope axono

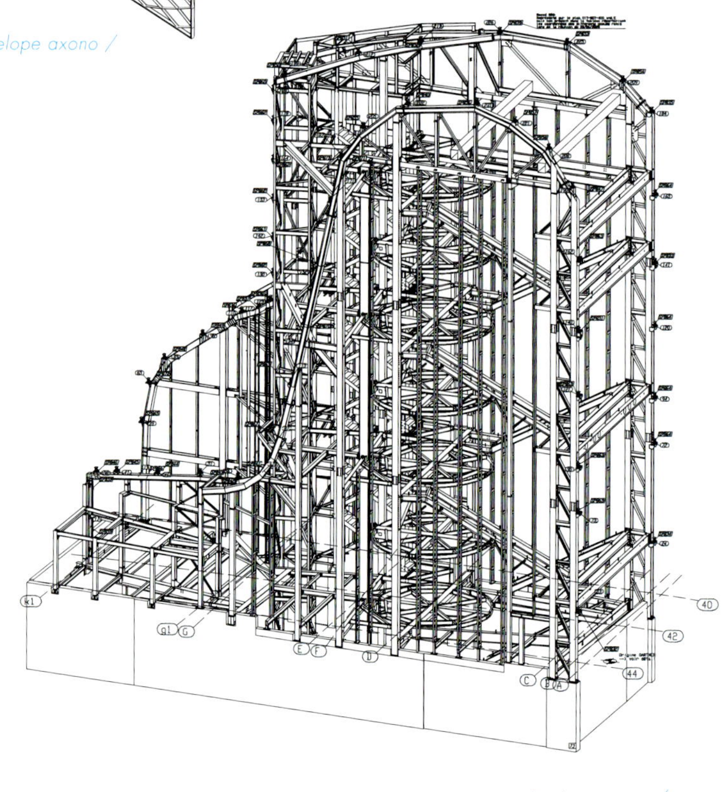

structure axono /

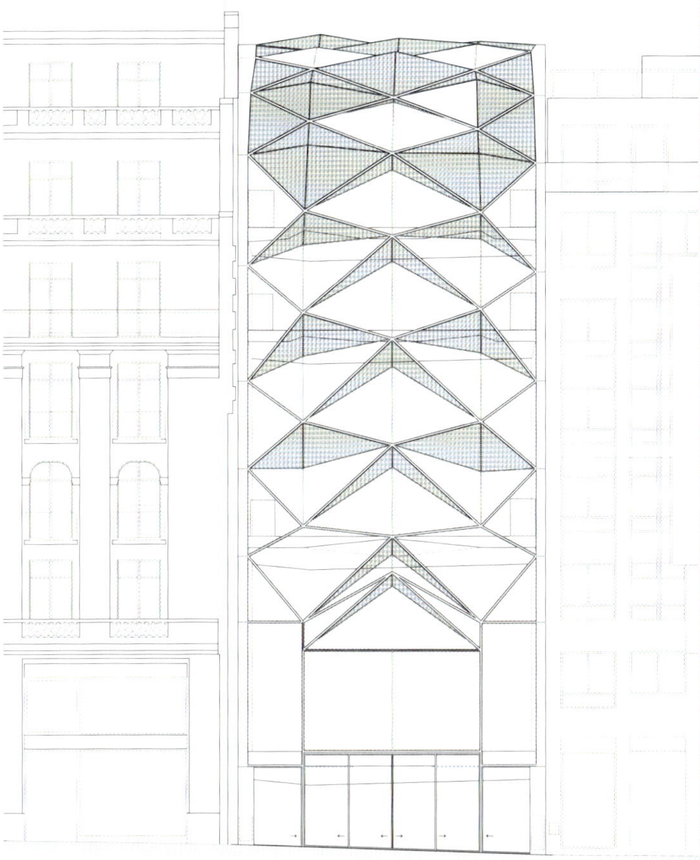

\ facade

ZARA

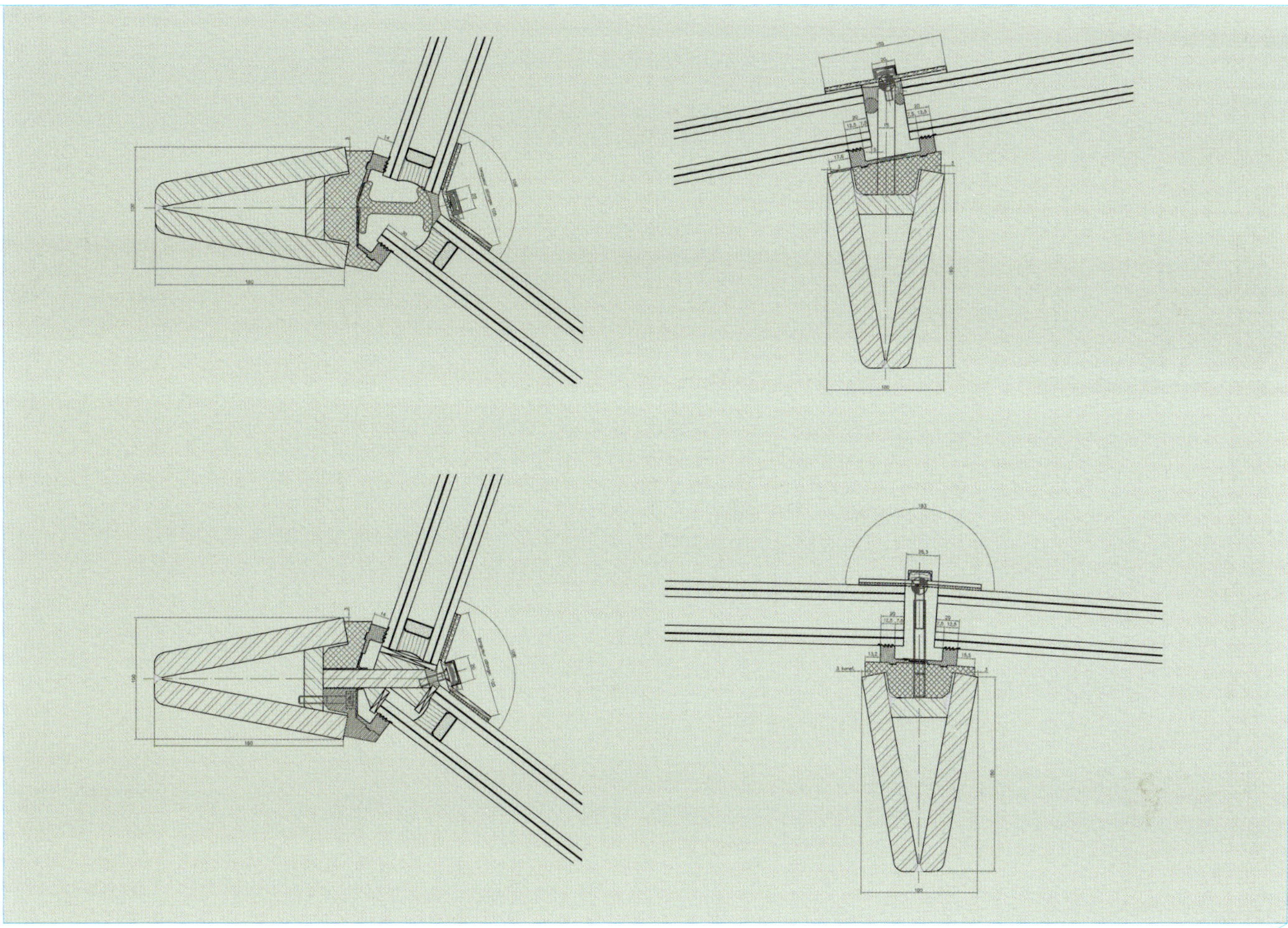

detail resille arriere EXE

detail resille avant EXE

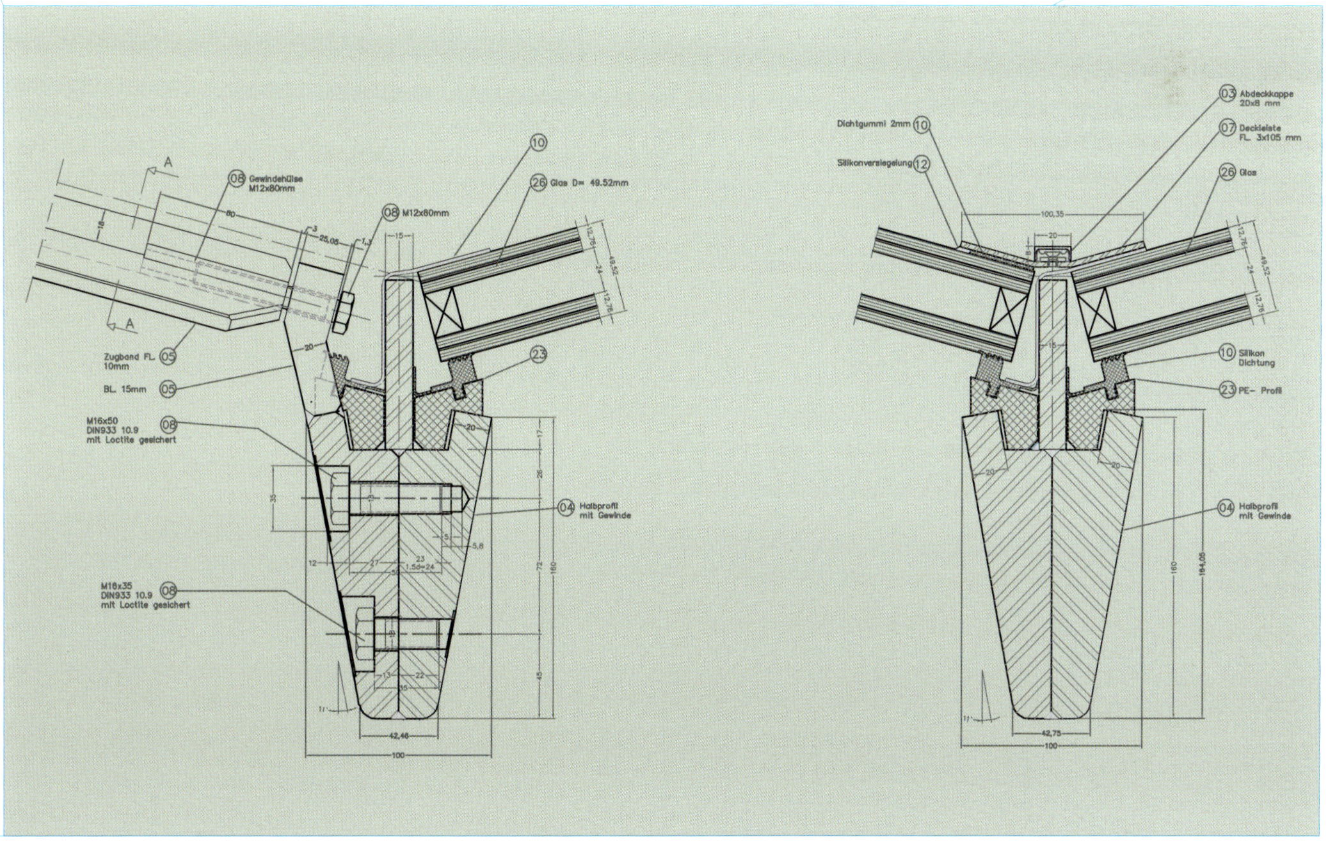

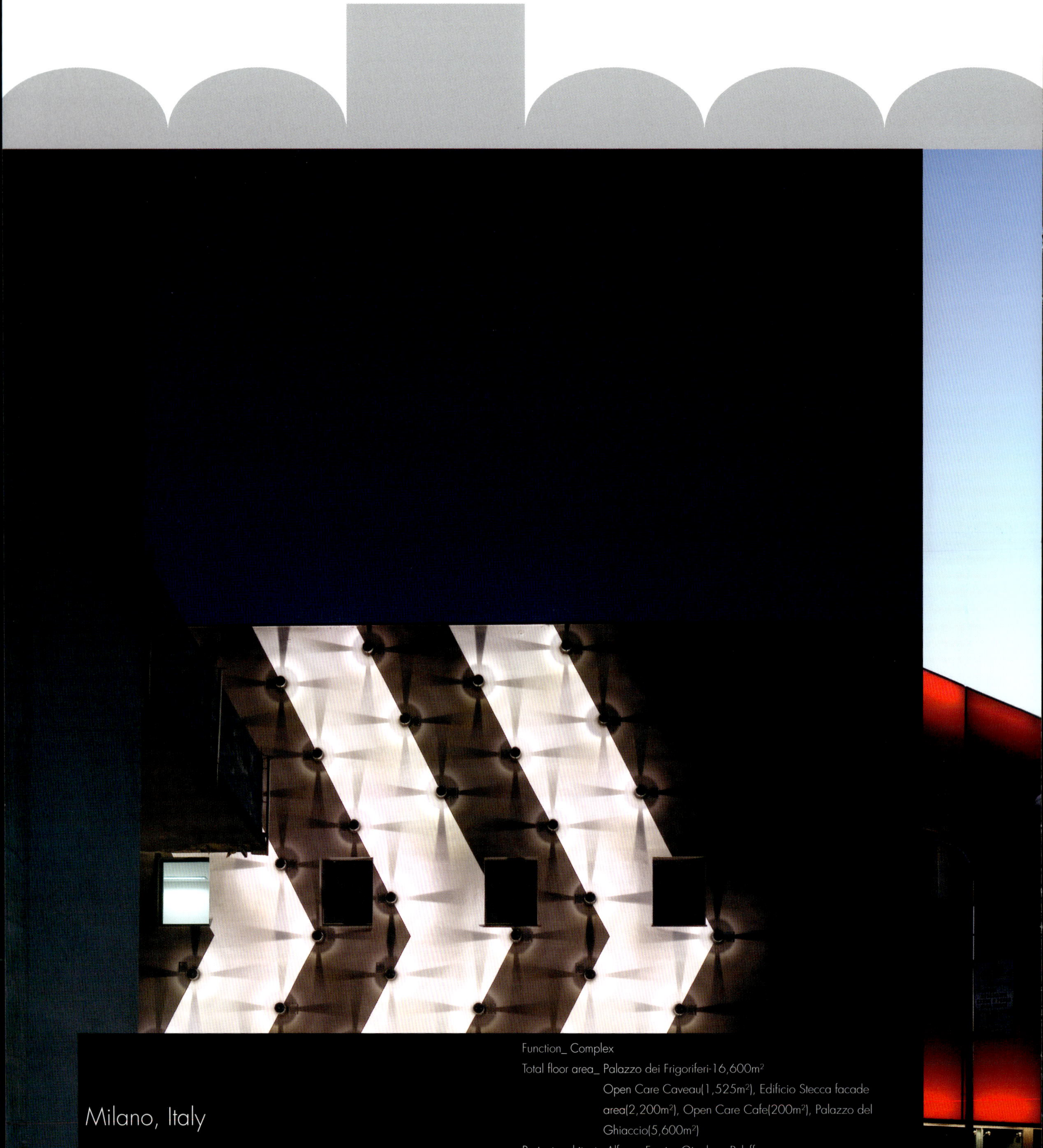

Milano, Italy

Function_ Complex

Total floor area_ Palazzo dei Frigoriferi-16,600m²
Open Care Caveau(1,525m²), Edificio Stecca facade area(2,200m²), Open Care Cafe(200m²), Palazzo del Ghiaccio(5,600m²)

FRIGORIFERI MILANESI AREA

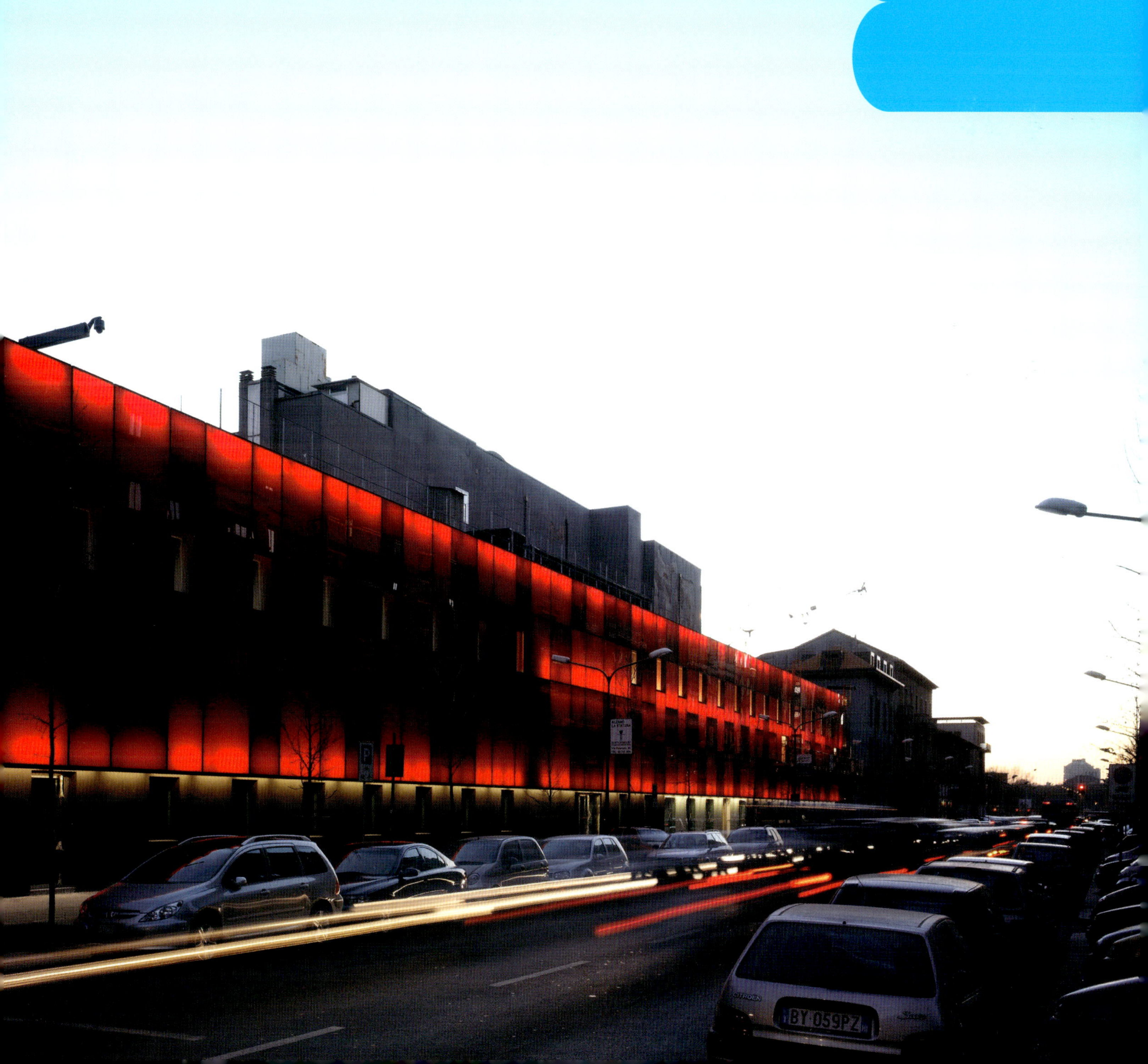

새롭게 들어서는 빈 공간, 그리고 그에 대한 해석되고, 조정되고, 강조된 모순으로 포르타 비토리아(Porta Vittoria)라 불리는 밀라노 내의 새로운 구역이 활기를 찾아가고 있다. 이 빈 공간의 옆면으로 길게 밀집된, 대부분 '생산기능'을 가진 전면부로 특징지어진 거리인 비아 피라네시(via Piranesi)와 균등하게 나누어진 '거주기능'을 지닌 또 하나의 거리를 볼 수 있다. 옛 철도의 선로는 두 전면부를 분리하는 긴 선형의 구획을 만들어내고, 덕분에 비아 피라네시를 내다보는 남쪽 영역과 포르타 비토리아를 바라보는 혼란스럽지만 매력적인 후면부가 만들어진다.

이 영역 내에서 '프리고리페리 밀라네시-팔라쪼 델 기아치오(Frigoriferi Milanesi-Palazzo del Ghiaccio)' 콤플렉스는 주도적인 역할을 한다. 콤플렉스는 비아 피라네시(Via Piranesi) 주변에 있는 길고 짧은 건물들로 이루어져 있고, 평범하고 단조로운 건물이지만 역사적 가치가 있는 두 건물, '팔라쪼 델 기아치오(Palazzo del Ghiaccio)'와 현재의 모습에서 아직도 회벽 너머로 그 흔적을 찾아볼 수 있는 독특한 특징을 많이 상실한 거대한 '모놀리스(monolith)'인 팔라쪼 데이 프리고리페리 밀라네시(Palazzo dei Frigoriferi Milanesi)에 대한 시각적인 관점에서 주요한 역할을 담당한다. 전체 콤플렉스의 회색은 밀라노와 같은 도시의 전형적인 분위기와 잘 들어맞는다. 중요한 지하 공간인 프리고리페리 밀라네시의 카보(Caveau)는 언급할만하다.

프리고리페리 밀라네시의 카보는 너무나 다르지만 언제나 대화를 추구하는, 매력적이고 혼란스러운, 이상한 볼륨과 기능의 집합체이다.

이번 복원 프로젝트는 이런 모든 면과 모순들에 초점을 맞추고 특히, 지역과 조화를 이루는 예술품과 귀중품의 주제들을 다루고 있다.

도시적 양상은 새로운 무역과 서비스 전략과 함께 혼재되어 있다.

이 프로젝트는 오픈 케어 브랜드와 새로운 이미지의 창조로 완결되며, 프로젝트의 출발점 중 하나가 된다. 그런 변화의 첫 단계로 비아 피라네시 속의 에디피치오 스테카(Edificio Stecca)와 오픈 케어 카보와 접근 경사로에 초점을 맞추고 있다.

개성을 상실한 낮고 긴 건물은 간단한 두 가지 프로세스에 의해 강조될 것이다. 그 프로세스 중 하나는 검은색 페인트의 코팅이 구성적 추론의 모든 형식의 흔적을 없애고 구조물을 모든 것을 삼켜버리는 블랙홀로 전환하기 위해 벽체, 창문, 윤곽 등 모든 것을 감추는 얼룩인 회색을 대신할 것이고, 새로 유리로 설치되는 광택 나는 '외피'는 주변 지역의 단조로운 기하학에 대한 다채로운 효과를 만들어내도록 중첩되어 이차원적인 '야간'효과를 통해 그 덩어리의 경계를 허문다.

이 프로젝트는 구조체에 새로운 일체화된 형상을 강화하고자 의도된 양쪽 파사드 모두에 과도한 크기의 그래픽을 적용함으로써 (프리고리페리 밀라네시, 이러한 형상은 건축주의 의도에 따라 시공 중에 상실되고 말았다.) 외부의 공공영역과 사적인 내부공간 사이에 '투명한' 느낌을 부여하기 위한 새롭게 깊숙한 지하개방공간을 만들어 완성한다. 이 프로젝트와 구조체의 볼륨은 전체적으로 디테일에 대한 연구가 부족함을 드러낸다.

건물은 이중의 미학적 측면으로 도시의 '랜드마크'가 되었다(도시를 향해 부유하는 색채를 지닌 건물이자, 프리고리페리 밀라네시 속에 있는 일체식의 회화적 건물이다). 100미터나 되는 길이 덕분에 투시도적 효과를 내며, 그 지역의 크고 밝은 화살표와 같은 출입구를 가리키고, 도시와 함께 그 지역에서 벌어지는 활동들을 연결하는 커뮤니케이션 요소인 '플러그 빌딩'이 된다.

에디피치오 스테카를 감싸는 검은 얼룩은 카보의 진입경사로 위로 흘러나오며, 거기에서는 흰색의 붓과 밝은 선으로 그려진 얼룩말의 무늬와 유사한 장식 덕분에 어떤 새롭고 모호한 차원을 경험할 수 있다. 검은색은 방문자들을 현관홀 안으로 유인하는데 그곳은 주차장 램프와 미로 같은 통로와 출입문 사이에 있어서 여유롭고 물리적인 침묵으로 방문자들을 감싸는 공간이다.

세 개의 기둥으로 지지되는 사각의 현관홀은 여러 영역과 끊임없는 연관성을 만들어내면서 그 구조체를 지우고 분할한다. 뒤에 남겨진 모든 것은 짙은 검은색이다. 그와 달리 전면에 드러난 모든 것은 왼쪽으로 밝게 반사되고, 오른쪽으로 반투명 색이 입혀져 있다. 실의 중심에는 C자 모양의 흰색으로 칠한 오크 목재가 무형의 흰색 플랩 에드라 4개에 의해 강조되면서 현관홀의 윤곽을 만든다. 재료와 디테일에 대한 탐구는 엷은 것에서부터 짙은 색까지 여러 색채의 얼룩이 각 각의 방으로 연결된 복도를 나타내는 출입문 너머에 감춰져 있다.

풍부함과 필수적인 것, 보안과 커뮤니케이션의 사이에서, 우리는 소개하는 사람과 사물에 따라 규정하지 않음으로써 매혹적이고, 대중적이고, 우아하고, 냉정하고, 온화한... 오픈 케어 룸을 찾고 있다.

오픈 케어 카페는 이중적인 미학적 차원, 즉, 도시를 향한 다채로운 파사드와 프리고리페리 밀라네시 내에서 일체식이면서도 회화적인 정면을 통해 낡고 신원불명의 건물을 도시의 '랜드마크'로 바꾼 최근의 재개발 프로젝트가 벌어진 비아 피라네시르 바라보는 에디피치오 스테카(플러그 빌딩)의 지하에 놓여있다.

오픈 케어 카페는 도시와 오픈 케어 세계 사이의 중간적이고 '공공적인' 구역인 지하의 '투명한' 영역에 위치한다. 구조물은 고르지 않은 천정고를 감추면서, 가득 차고 빈 덩어리, 집중되면서 확장된, 도시적이면서 가정적인 공간으로의 변화가 특징적이다.

공간을 형성하는 기하학(길고, 좁고, 높음) 덕분에 오픈 케어 카페는 그 공간을 다른 방식으로 즐길 수 있는 장소인 두 개의 영역을 구별한다.

나무 '카펫'은 벽난로가 있는 느리고 가정집과 같은 한 측면과 나무, 철, 유리로 이루어진 카운터가 있는 빠르고 도시적인 다른 한 측면을

A new area is coming to life in Milan: the area of Porta Vittoria, together with its new empty spaces to fill in and its contradictions to interpret, handle, emphasize. On the sides of these empty spaces we find via Piranesi, a street which is characterized by one long and compact front mainly 'productive' and another equally fragmented, but 'residential'. The old railway track creates a linear zoning, separating the two fronts thus generating the south front overlooking Via Piranesi and the chaotic but charming back area overlooking Porta Vittoria.

Within this area the complex of Frigoriferi Milanesi-Palazzo del Ghiaccio has a predominant role. It is composed of a short and long building, located along Via Piranesi. It is a banal and dull building, but it plays a main role from the perspective point of view with respect to the other two important buildings beyond the line, Palazzo del Ghiaccio, a historic building, and Palazzo dei Frigoriferi Milanesi, a massive 'monolith', which in its current physiognomy has lost many of its distinctive characteristics, still traceable beyond the plastered wall. The grey colour of the whole complex perfectly matches with the typical atmosphere of a city such as Milan. Worth mentioning an important underground basement: Frigoriferi Milanesi's Caveau, which makes the eccentric set of volumes and functions, so different but always in search of a dialogue, even more charming and confusing.

The restoration project focuses on all these aspects and their contradictions and deals with the themes of art and treasures - particularly in tune with the area.

The urban aspects mix up with new trade and services strategies; the project is completed by the creation of the Open Care brand and a new image: one of the starting points of the project. The first steps of such transformation focus on the Edificio Stecca in Via Piranesi, on the Open Care Caveau and its access ramp.

The lack of personality of the long and low building will be highlighted by two simple processes: on the one hand a coat of black paint will replace the grey colour, a blob which assails everything: walls, windows, profiles, in order to obliterate any form of architectonic ratiocination and to transform the structure into a black hole, swallowing everything; on the other a new shiny 'skin-peel', made of glass, will be overlapped in order to create a chromatic effect with respect to the monotone geometry of the surrounding area; it unedges the volume with a bidimensional and 'night' effect.

The project is completed by the realization of new and deep openings in the basement in order to confer a feeling of 'transparency' between the outside public area and the private interior, by the graphical over dimensioning (Frigoriferi Milanesi) on both facades (this aspect was lost during the works according to the commissioner's will) intended to enhance the new monolithic appearance of the structure and by the lighting system realized in Via Piranesi and in the inner courtyard. The project and the volume of the structure are marked by a total lack of research for details.

The building has become an urban 'landmark' with a double aesthetic dimension (a coloured one suspended towards the city, and a monolithic and graphic one within Frigoriferi Milanesi). Thanks to its 100 meters length, it plays with perspective and it points out as a big, bright arrow the entrance to the area, becoming a 'Plug Building': an element of communication, connecting the activities taking place inside the area with the city.

The black blob covering the Edificio Stecca oozes over the access ramp of the caveau, where it is possible to perceive a new ambiguous dimension thanks to a zebra-striped decoration made of white brush-strokes and bright lines. The black colour drives the visitors inside the foyer, where a room, suspended between a garage ramp and a labyrinth of passages and doors, surrounds them with a relaxing and physical silence.

The foyer, a rectangle supported by three pillars, cancel and divides its structure, creating different areas and continuous connections. All that is left behind is black, deep. All that stands in front is different: bright, curved and reflecting on the left, semi-transparent and coloured on the right. At the centre of the room, a C-shaped peace of whitened oak, defines the foyer, punctuated by four white and shapeless Flap Edra. The concentration of materials and details is obliterated beyond the door, where blobs of different colours, from light to dark, characterise the corridors accessing to the cells.

Between richness and essentiality, safety and communication we are looking for space without defining it, so that according to whom and what is being introduced, can become glamorous, pop, elegant, cold, welcoming... the Open Care room.

The Open Care Cafe is located in the basement of the Edificio Stecca (Plug Building) overlooking via Piranesi, where a recent requalification project changed a shabby and anonymous building into a urban 'landmark', with a double aesthetic dimension: a coloured facade towards the city, and a monolithic and graphic frontage within Frigoriferi Milanesi.

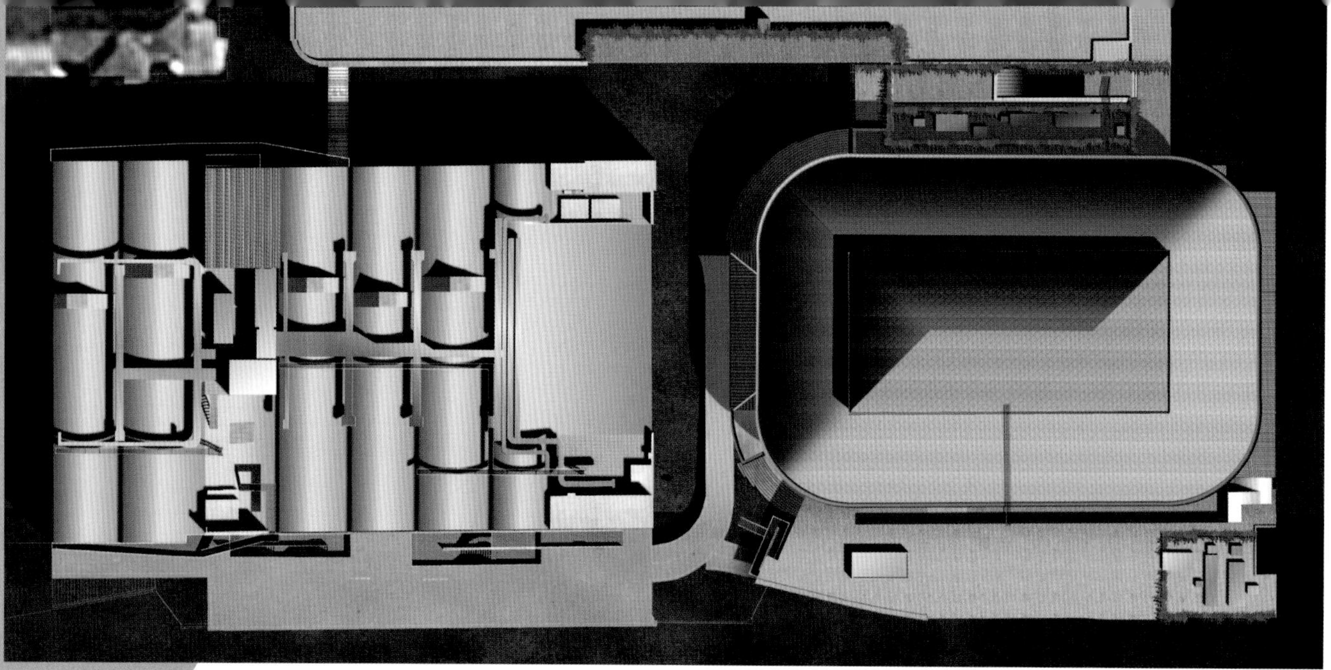

통해 두 개의 다른 장소와 양상을 대비시키면서 식당 영역을 수직, 수평으로 둘러쌌다. 흰색이며 광택이 나는 두 개의 띠로 된 레진은 오픈 케어 카페를 가로와 외부 조경으로부터 구분하면서, 공간을 가로지르는 규칙을 나타낸다.

공간은 두 개의 '상자'로 이루어져 있는데 외부에서도 들여다볼 수 있는 투명한 상자가 그 하나이고, 오픈 케어 카페의 고객 모두에게 일종의 스카이라인이라고 할 수 있는 빛으로 이루어진 표피에 의해 한정된 흰색의 박스가 또 다른 하나이다.

프리고리페리 밀라네시(Frigoriferi Milanesi). **모놀리스**(The monolith)
둔탁하면서 강력한 도시적 아이템

기능의 다양성 즉, 오픈 케어-프리고리페리 밀라네시의 행위들 덕분에 비아 피라네시를 내다보는 한쪽 면인 후면 구역은 남부 지역 내의 주요 파사드가 된다.

프리고리페리 밀라네시 프로젝트는 기존의 규모를 확장하고, 도시와 그 영역에 대해 활발한 역할을 수행하는 도시적 공간으로 전환시키는 것에 집중한다.

모든 것은 피라네시를 통해 들어가서 거의 한데 합쳐진 이질적인 건물들로 이루어진 세계 속에 있음을 발견하게 되는 사람들의 움직임에 의해 생성된다.

도시와 '극도로 보안된' 지역 사이의 경계선을 건너간 후 뒤에 남겨진 거리는 이 모순적인 장소와 자극적인 힘의 충만함을 모호하고 매력적으로 만드는 기반시설적인 특징과 함께 우리를 인도한다(카보를 향해, 팔라쪼 데이 프리고리페리의 남측 출입구를 향해 사물들을 끼워 넣는 팔라쪼 델 기아치오와 계단 시스템 말한다). 건물은 공제행위(틈-마지막 층의 개방공간)를 통해 그 본질과 형상, 그리고 부가적인 움직임(4층으로 연결되는 발코니와 테라스가 설치된 산책로-계단으로 이루어진 시스템)을 다시 얻게 될 것이다.

건물은 저층부의 3개 레벨에서 이루어질 감시기능과 아트 서비스를 통해 그것의 내향적인 성격을 유지하게 될 것이다. 우리는 마지막 층이라는 위치 덕분에 오픈 케어 스페이스와 더불어 주요 예술 행사를 수용하고 도시와의 관계를 발전시켜 나가게 될 공공영역 등의 새로운 장소를 발견하게 될 것이다.

아직 존재하지 않는 한 장소는 숨겨져 있지만, 자체적인 형상과 역할을 가질 준비가 되어있다.

우리는 선박 건조 기술로 만들어진 20미터 깊이의 빛나는 틈을 따라 내려감으로써 건물의 내부로 내려가 그 '실체'를 발견할 수 있다.

로프, 도르레, 균형추를 활용해 최상층까지 운행되는 가볍고 투명한 캐빈 시스템을 통해 다시 위로 올라갈 수 있을 것이고, 최상층은 자연광과 인공조명에 의해 조절되며, 모든 방향으로 개방된 특별한 공간을 발견하게 될 것이다.

신기술 덕분에 실현된 베네치아 양식의, 일체형이며 단색으로 된 평면은 공간적 범위를 증가시킬 것이다. 건물을 위해 특별히 설계된 조명 시스템은 천장 시스템을 강조한다. 새로 설치된 천창은 해시계처럼 동쪽에서 서쪽으로 삽입되어 시간을 나타내게 될 것이다. 대부분 전통적인 재료가 사용될 것이고, 가득 찬 공간과 빈 공간 사이, 회반죽 요소와 철골 간의 균형, 그리고 불투명성과 맑고 투명함 간의 균형에 초점을 맞추게 될 것이다. 이것은 밀라노를 위한 새로운 도전이고 새로운 건물이다.

팔라쪼 데이 프리고리페리 밀라네시(밀란 얼음 궁전)은 얼음의 생산과 냉동고 속의 창고를 위한 공동체로서 1899년 설립되었다. 팔라쪼 델 기아치오(얼음 궁전)는 1923년 건설되었다. 그것은 오랫동안 유럽에서 가장 큰 실내 아이스 스케이팅 링크였다. 1950년대부터 그 기업은 냉동고를 보석 및 귀중품을 보관하기 위해 보안 관리된 공간으로 변경할 것을 제시했다. 그리고 회화, 골동품, 도자기, 조각, 태피스트리, 카페트 등을 보존하기 위한 특수한 보안 및 공조 시스템을 도입했다.

2003년 새로운 전략이 적용되기 시작했다. 일반적으로 복원전문가, 경매자, 단독 상담가, 비전문적인 운송업자 등 다양한 운영자들에 의한 모든 아트 서비스가 한 특별한 운영자 아래에 통합되었다. 오픈 케어 프리고리페리 밀라네시는 예술 작품의 보관을 위한 어떤 요구사항도 충족하는 유일한 매개자가 되기를 원한다.

Open Care Cafe is located in the 'transparent' area of the basement, an intermediate and 'public' area between the city and the Open Care world. The structure is characterized by the alternation of full and empty volumes, compressed and expanded, urban and domestic spaces, hiding the uneven heights of the ceilings.
Thanks to its geometry that shapes the space(long, narrow, high), the Open Care Cafe identifies two territories where it is possible to enjoy the space in different ways.
A wood 'carpet' vertically and horizontally enclosed the food area, putting in contrast two different places and attitudes: one slow and homelike, with a fireplace; the other fast and urban with a counter made of wood, steel and glass. Two strips of white and glossy resin define the rules for crossing the space, separating the Open Care Cafe from the road and the outside landscape.
The space is composed by two 'boxes': a transparent one, also visible from the outside, and a white one, defined by surfaces made of light, a sort of skyline for every Open Care Cafe's customer. The sky is enriched by Bubble-Norlight and the sittings by Sottssas for Segis(Trono). Open Care Cafe is the new 'epi-centrum' of Milan, it is the first entrance to a different world, anticipating the great transformation of Palazzo dei Frigoriferi and Palazzo del Ghiaccio.

Frigoriferi Milanesi. The monolith.
A dull and powerful urban item.

The back zone becomes the main facade within the southern area, a front overlooking Via Piranesi, intended to be a 'skyline', thank to the diversification of its functions: the Open Care- Frigoriferi Milanesi's activities.
The project of Frigoriferi Milanesi focuses on enhancing the existent volume and on transforming the building into an urban space playing an active role for the city and the territory.
Everything is generated by the movements of the people who, entering via Piranesi 10, find themselves in a world composed of heterogeneous, almost merged together buildings.
The street, which we left behind after having crossed the border line between the city and the 'high security' area, carries on leading us, together with its infrastructure character, making this contradictory place and full of evocative forces, ambiguous and attractive (plugging object, Palazzo del Ghiaccio, the staircase system, towards the Caveau, towards the southern entrance of Palazzo dei Frigoriferi).
The building will regain its nature and its shape through a subtraction action (the gap, the opening on the last floor) and an addition action (a system of promenade-stairs equipped with balconies and terraces, leading to the fourth floor).
The building will keep its introvert nature thanks to the supervision activities and the art services that will take place within the first three levels. We will discover a new place: the Open Care Space, a public area that will house the main art events and will develop its relationship with the city, thanks to its location on the last floor.
A place that does not exist yet, that is hidden, but ready to take on its own shape and role.
We can go down the internal part of the building and discover its "body", by descending along a 20 metres deep and shiny gap, built with a naval technology. It will be possible to go up again by a light and transparent system of cabins lifting us up to the last floor, by means of ropes, pulleys, counterbalances… on the last floor we will discover a unique space, opened towards all directions, modulated by the natural and artificial lights.
A monolithic and monochrome floor, in Venetian style, realized thanks to a new technology, will strengthen the spatial dimension. A lighting system specifically designed for the building will highlight the vaulting system. New skylights will be inserted and placed from east to west like a sundial, will indicate the time. Most of the materials used will be traditional, focusing on the balance between full and empty spaces, plaster elements and steel profiles, between opacity, brightness and transparency. A new challenge and a new building for Milan.
1 Palazzo dei Frigoriferi Milanesi (Milan Fridges Palace) was founded in 1899 as a society for the production of ice and for the storage in refrigeration cells. Palazzo del Ghiaccio (Ice Palace) was built in 1923. For a long time it remained the biggest indoor ice skating rink in Europe. From the '50s the company changed the refrigeration cells into armoured and conditioned rooms to keep jewellery and valuable objects. Then the company adopted specific security and air conditioning systems for the conservation of paintings, antique objects, pottery, sculptures, tapestries and carpets.
In 2003 the new strategy has been started, all the art services generally managed by different operators such as restorers, auctioneers, single advisors, not-specialized transporters, have been united under a unique operator. Open Care Frogoriferi Milanesi wants to be the unique intermediary to comply with any kind of request for the custody of art works.

5+1AA

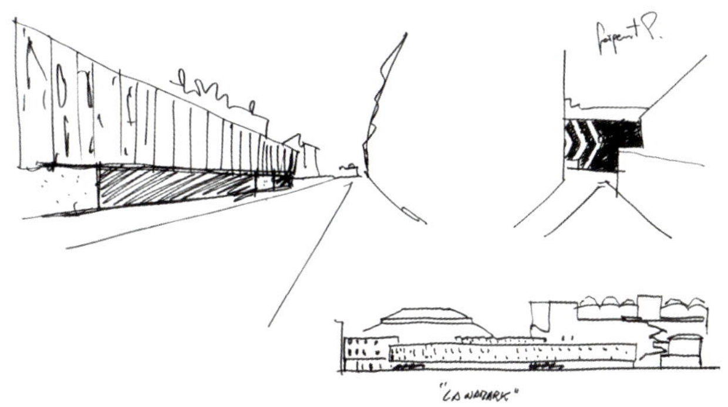

Edificio Stecca "Plug Building" - sketch /

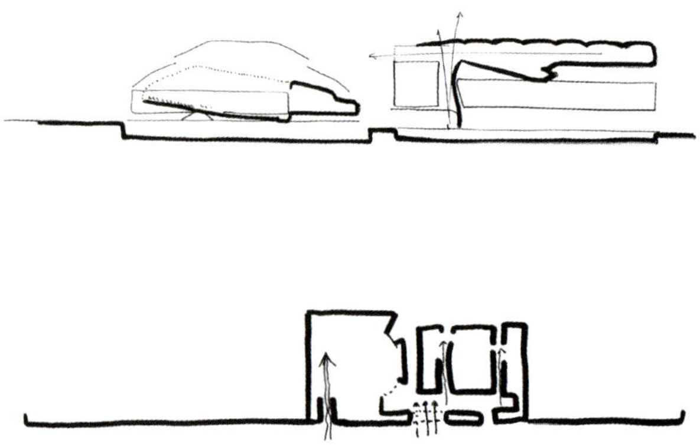

\ Frigoriferi Milanesi Area - distribution sketch

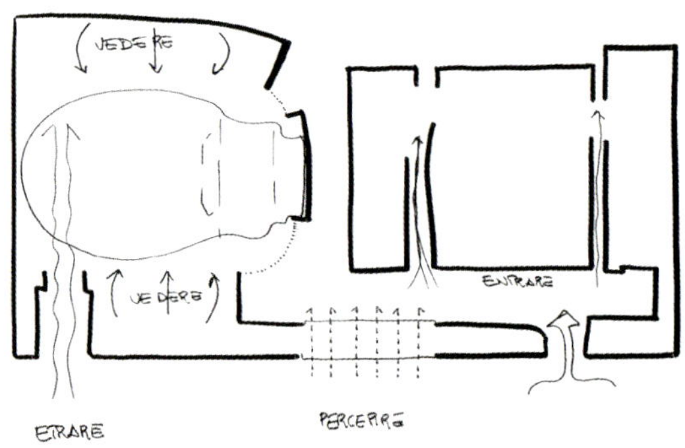

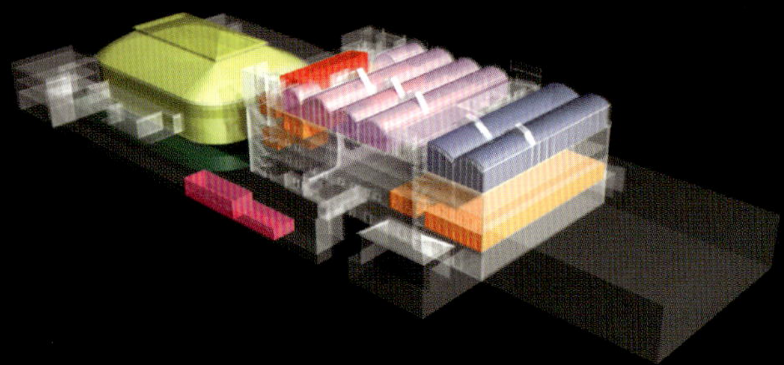

\ Total longitudinal section

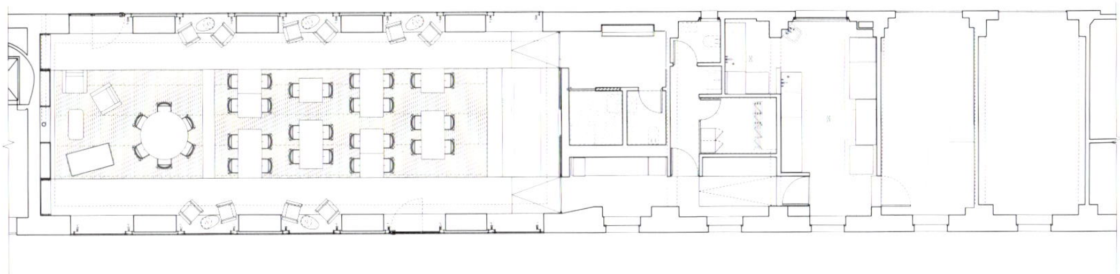

Open Care Cafe - floor plan /

Total floor plan \

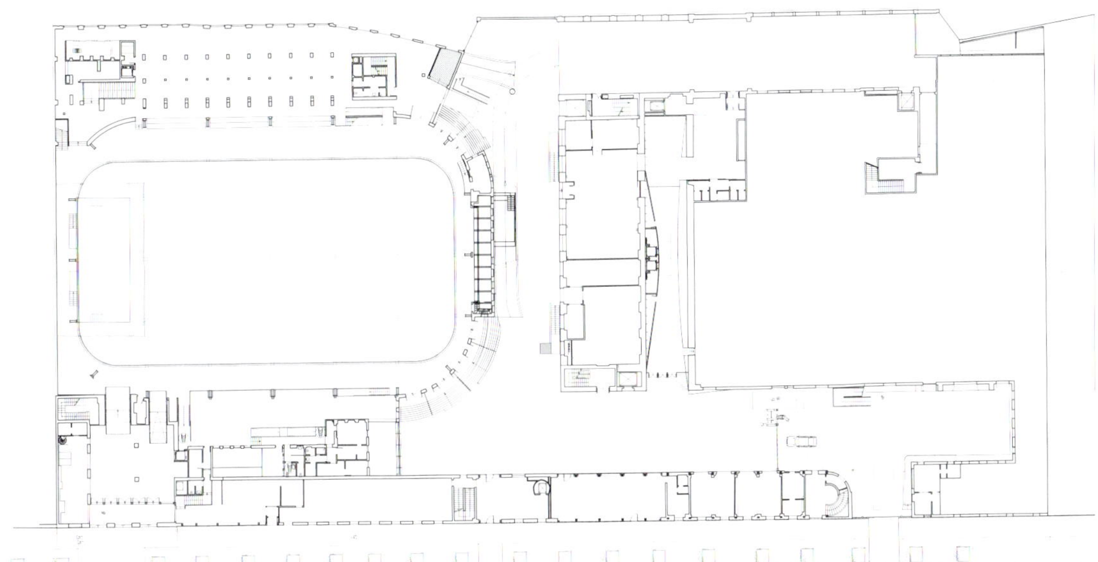

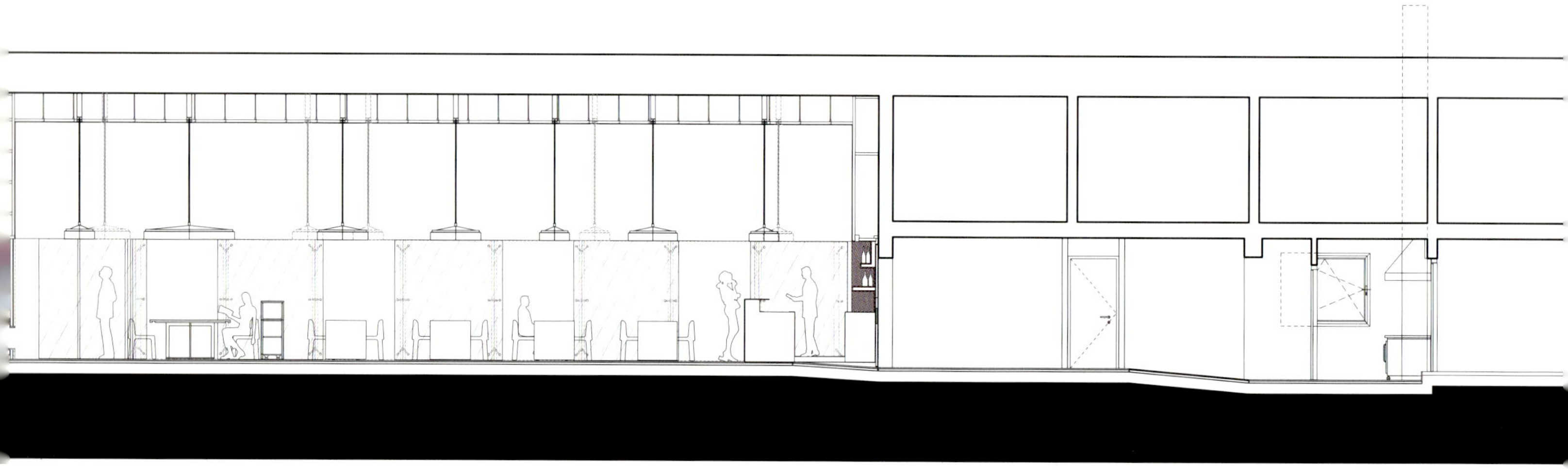

\ Open Care Cafe - section

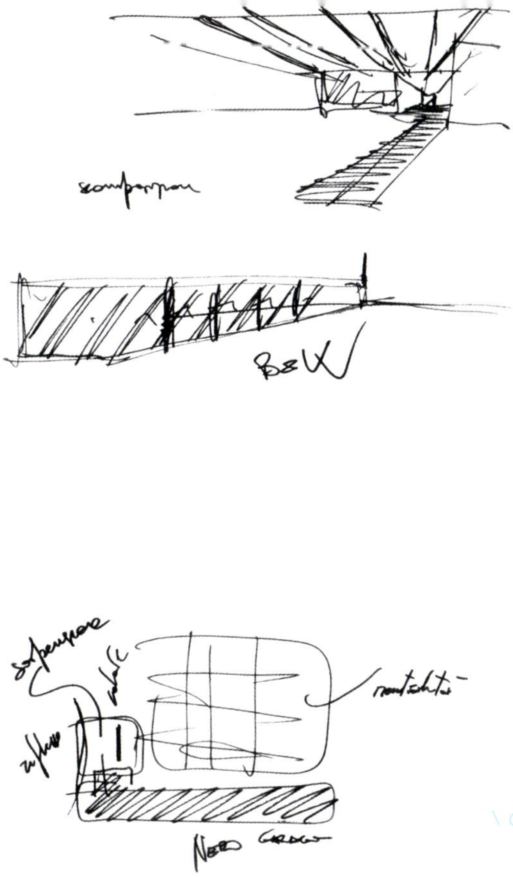

\ Open Care Caveau - sketch

Open Care Caveau - waiting room /

\ Open Care Caveau - safe deposit boxes

\ Open Care Caveau - access ramp

Open Care Caveau - floor plan /

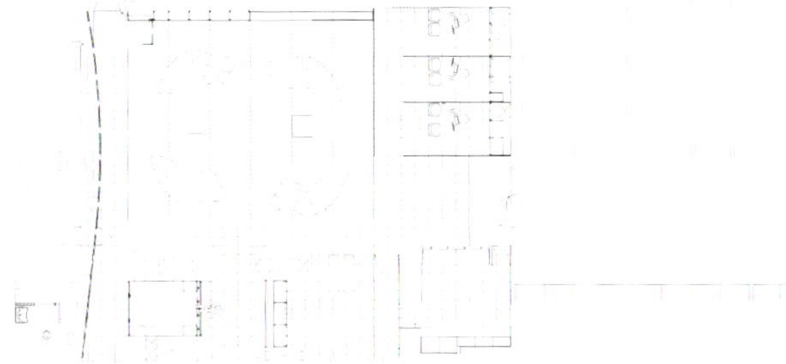

Open Care Caveau - curved wall /

\ Open Care Caveau - access ramp

Palazzo del Ghiaccio - clay model by Danilo Trogu /

\ Palazzo del Ghiaccio - longitudinal section

Palazzo del Ghiaccio - historical shot /

\ Palazzo del Ghiaccio - cross section

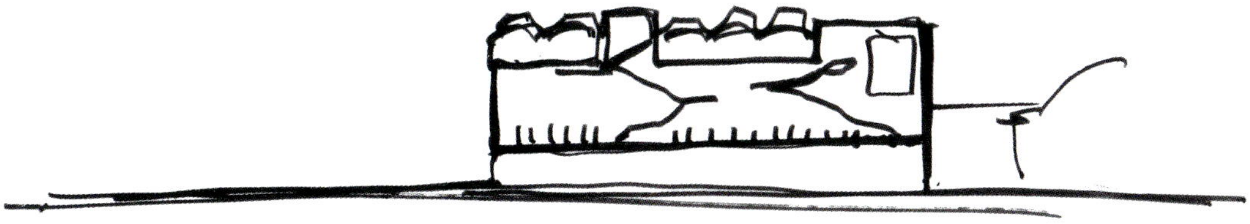

\ Palazzo dei Frigoriferi - sketches

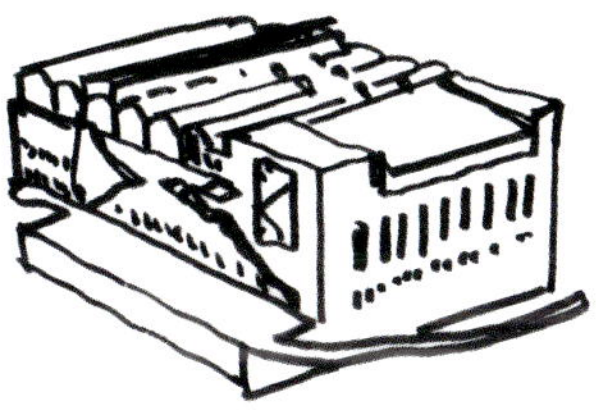

Palazzo dei Frigoriferi - clay model by Danilo Trogu /

Palazzo dei Frigoriferi - model with Throat /

\ Palazzo dei Frigoriferi - clay model by Danilo Trogu

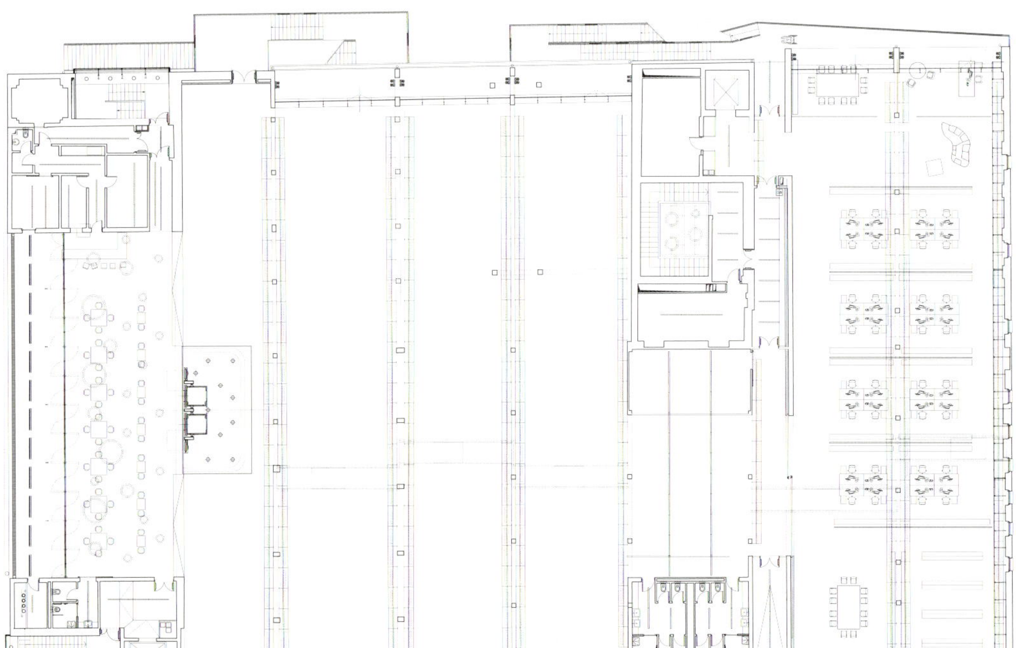

Palazzo dei Frigoriferi - 4th floor plan /

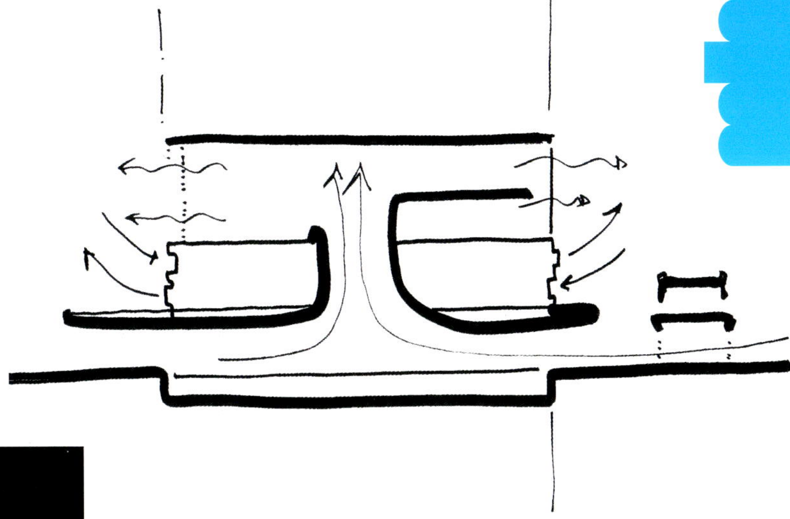

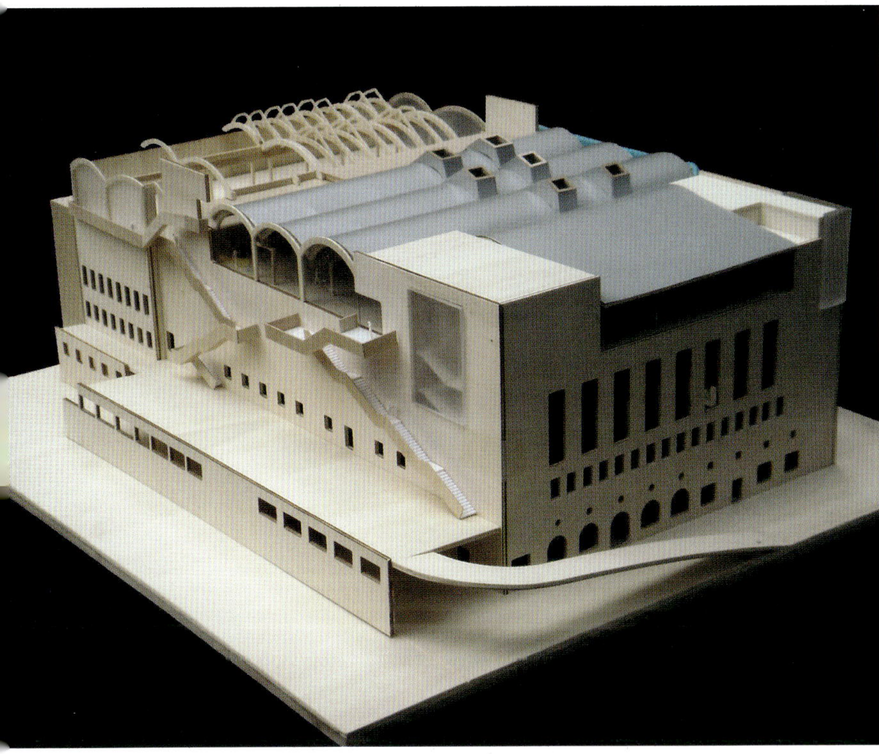

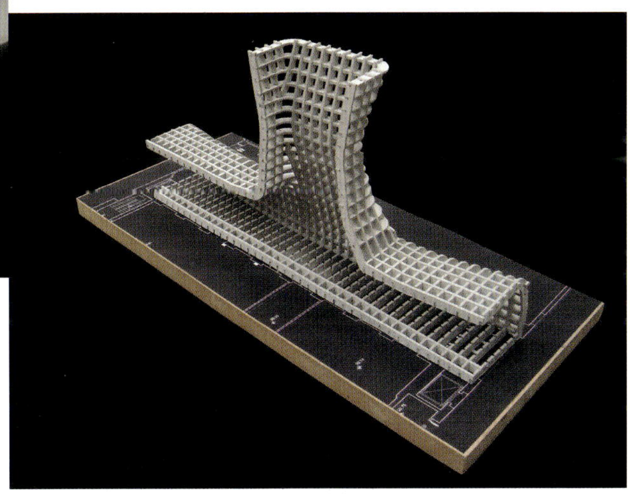

\ Palazzo dei Frigoriferi - Throat model

Palazzo dei Frigoriferi - Historical shot /

Leidsche Rijn Utrecht, Netherlands

2005

ONL [Oosterhuis_Lénárd]

Function_ Showroom, office
Total floor area_ 6,400m^2
Photographer_ ONL
Client_ Hessing Holding BV
Project architect_ Sander Boer, Kas Oosterhuis, Ilona Lénárd
Structural Engineer_ ONL, Meijers Staalbouw BV

HESSING COCKPIT IN ACOUSTIC BARRIER

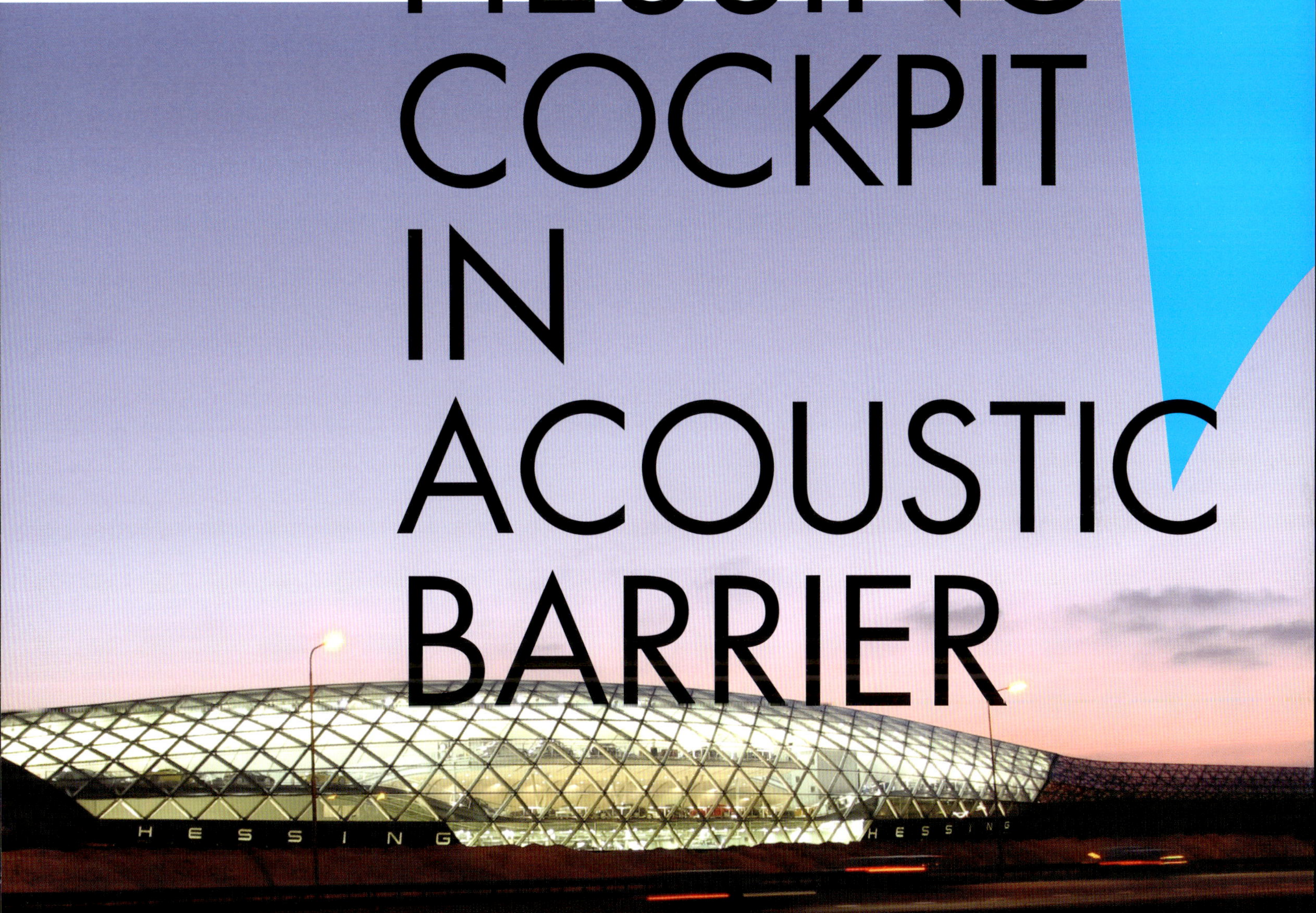

방음벽과 콕핏(조종공간)은 시속 120km의 속도로부터 생성된다. 자동차, 고속모터보트와 비행기는 마찰을 줄이고자 유선형으로 만들어진다. 건축물은 스스로 이동하지 않지만 자동차들이 지나는 연속적인 흐름을 따라서 위치한다. 건물 볼륨의 비율은 1.5km 길이로 길이와 높이의 비가 10:1인 비율에 따라 자동차가 지나가는 방향으로 뻗어나가 있다. 방음벽은 콕핏의 볼륨을 따라 부풀어있다. 콕핏은 [RR, Bentley]와 같은 고급 차량을 위한 쇼룸과 작업공간이 포함된다. 내부적으로 쇼룸의 평면은 닫힌 고리로 연결된 4개의 구분되는 레벨로, 유리 파사드의 비정형(NURBS) 곡선을 따라서 형성된다. 콕핏은 A2번 고속도로에서 15미터도 안 되는 거리에 위치한다. 유리지붕에서 유리파사드까지의 연속성으로 생성된 자연채광과 볼륨의 접근성 덕분에 자동차들은 고속도로에서 잘 보일 것이다. 주요한 디자인 원리는 방음벽 양쪽 면 위에 부드럽게 길게 뻗은 형태를 구성하는 연속적인 선형의 사용이다. 고속도로방향은 삼각형으로 된 특별한 유리 부재의 시스템으로 구축되며, 반대편의 정면은 개방형 금속 망사로 씌워진다. 방음벽의 길이를 따라서 길고 유연하게 뻗는 라인들은 분명한 시작과 끝이 없다. 단면에서 그 형태는 모나게 꺾어진 곳에서 부드럽게 꺾인 곡면을 따라서 볼록한 곳에서 오목한 커브로 변화한다. 스타일은 매스 맞춤식의 원리를 기반으로 한다. 반복되지 않는 각각의 요소는 고유한 길이와 인접한 요소와 상관되면서 생기는 고유한 각도를 가지고 있다. 전체 방음벽은 기본적으로는 3D 모델링에 의해서 형성되는 한 가지의 매개 변수적 디테일에서 형성된다. 두 번째 매개 변수적 디테일은 스틸 구축의 값과 콕핏의 A2파사드의 유리부재의 값을 나타낸다. 두 가지 디테일들은 일반적인 커브형태를 참조하고 있다.

콕핏과 방음벽은 공장 프로세스 과정의 엄격한 정렬에 따라서 생성된다. ONL은 3D모델의 표면으로부터 생성된 참조 점들의 수많은 지점들에 이르게 하는 스크립트(상세 설명서)를 작성해놓았다. 그 설명서는 모든 구성요소에 대한 3D 형상물을 만들고, 그들은 상관 데이터를 형성한다. 그것은 마치 깔끔하게 통제된 기계생산을 위한 표와 같다. 절단, 용접 그리고 드릴 기계들은 표에서 데이터를 읽고 모든 분화된 영역들에서 인간 노동의 간섭이 없이 생산한다. 모든 영역들에 고유 번호가 부여되고 빠르게 하나로 조여지는데, 우선은 시험용으로 실제크기의 모의 모델이 공장에서 만들어지고, 그 후에 건축 현장에서 직접 만들어진다. 건축과 시공은 이러한 방법으로 디자인 과정과 생산과정은 완전하게 연결되며, 최후의 가시적 결과물도 마찬가지다.

The Acoustic Barrier and the Cockpit have been designed from the speed of 120 km/hour. Cars, speedboats and planes have been streamlined to diminish drag. Buildings do not move themselves but are positioned along a continuous stream of cars passing by. The proportions of the built volumes inserted in the 1.5 km long stretched slender body the Acoustic Barrier are stretched 10-fold along in the longitudinal direction of the barrier. The barrier pumps up to host the volume of the Cockpit. The Cockpit houses a showroom and workplace for luxury cars [RR, Bentley]. Internally the floors of the showroom form a closed loop connection 4 discrete levels, following the NURBS curves of the glass facade. The Cockpit is positioned at no more then 15m distance from the side of the A2 highway. Thanks to the natural lighting produced by the continuity of the glass facade into the glass roof, and thanks to the proximity of the volume the cars will be extremely well visible from the highway.

The dominant design principle is the use of long continuous lines forming smoothly stretched shells on both sides of the barrier. The Highway side is built up with triangulated system of unique glass element, the other front side is clad with an open metal mesh. The long elastic lines along the length of the barrier have no explicit beginning or end. In the section the shapes change from angular to smoothly curved, from convex to concave curves. The styling is based on principle of mass-customization. There is no repetition, each element has an unique length and an unique angle in relation to the neighbouring element. The whole Acoustic Barrier basically is described in one single parametric detail, informed by the 3d model. A second parametric detail describes the values of the steel construction and the glass elements of the A2 facade of the Cockpit. Both details refer to a common reference curves.

Cockpit and Acoustic Barrier are produced according to a strict file to factory process. ONL has written scripts running on the point cloud of reference points which is extracted from the 3d surface model. The script produce the 3d visualizations of all constituting elements and they produce the relevant data - neatly administrated in a tabel ? for the production machines. The cutting, welding en drilling machines read the data from the table and produce without the interference of human labor all discrete parts. All parts are given a number and are quickly bolted together, first in the factory as a test model and full-size mock-up, later directly o the building site. Architecture and construction are in this way fully connected, both in the design process, the production process, and as a visual endresult. Architecture and construction have become one.

ONL

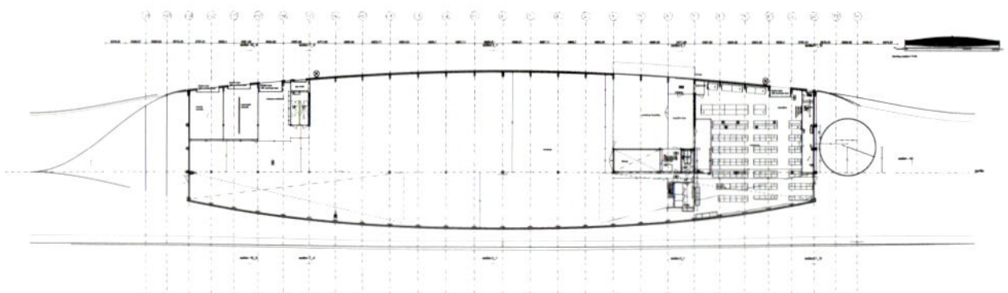

\ *level 0*

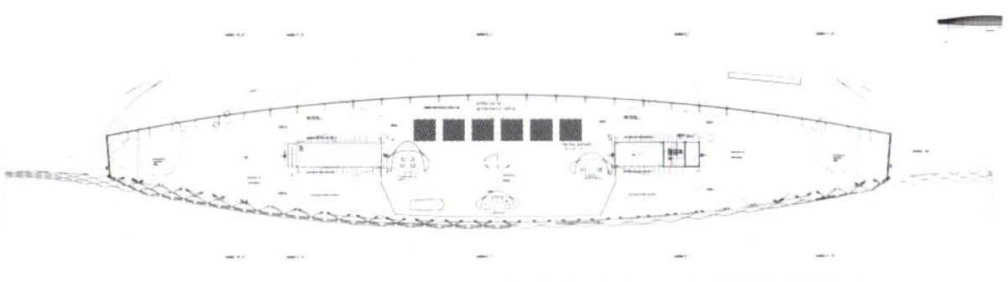

\ level 1

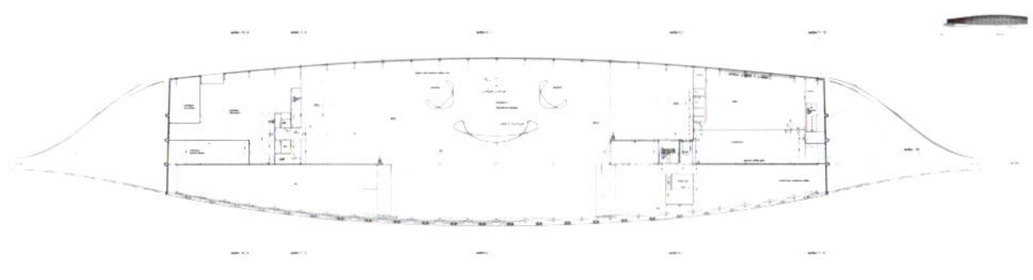

\ level 2

cross section /

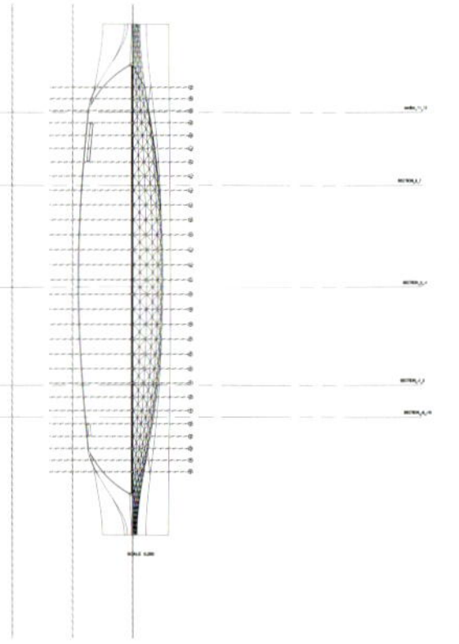

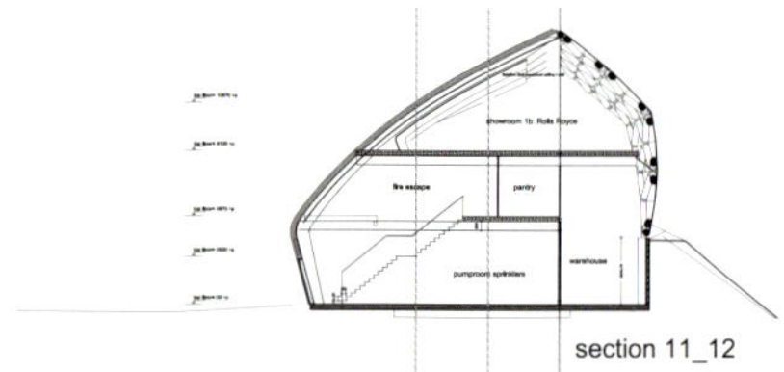

section 11_12

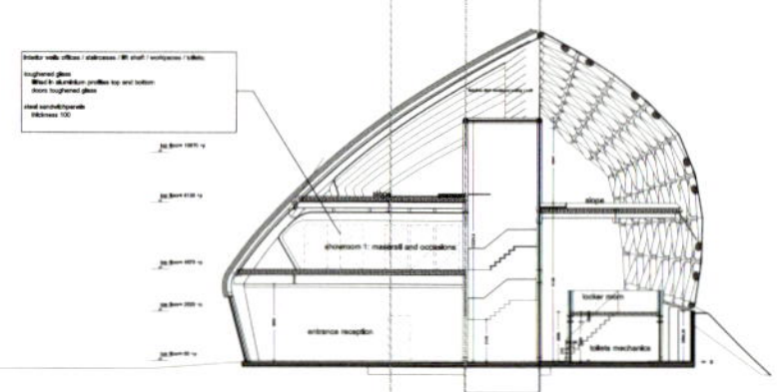

section 6_7

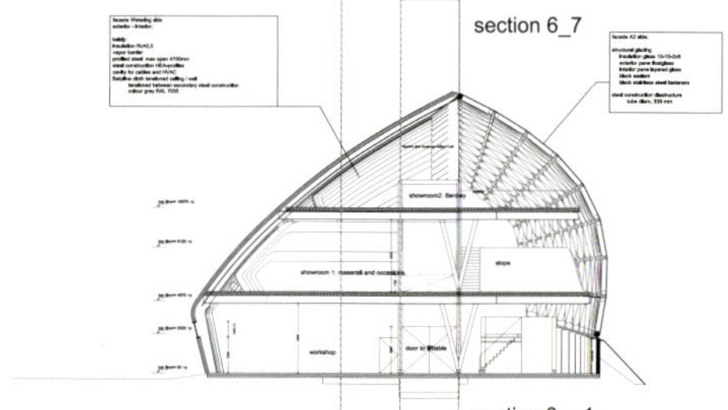

section 0 _-1

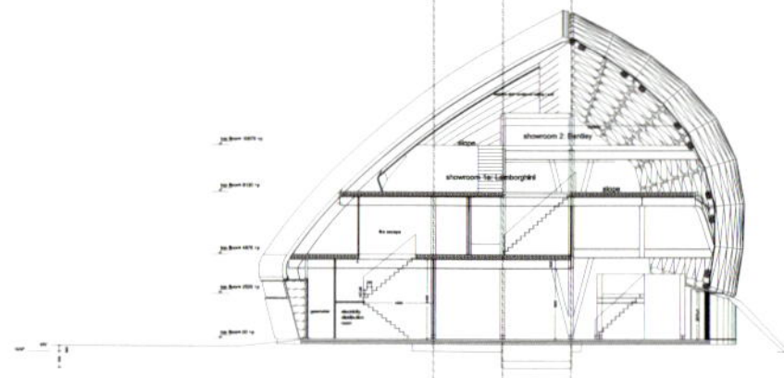

section -7 _ -8

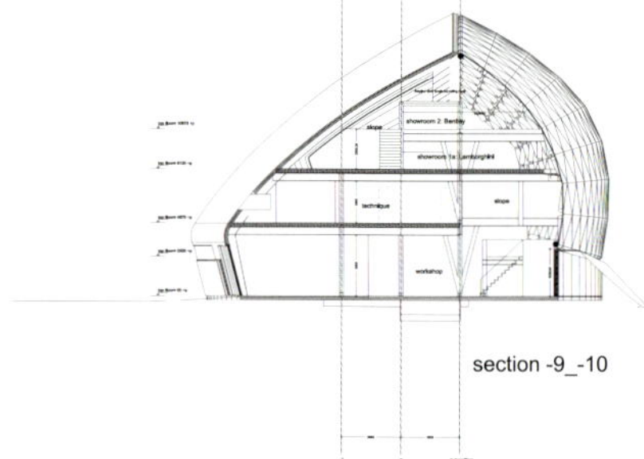

section -9_-10

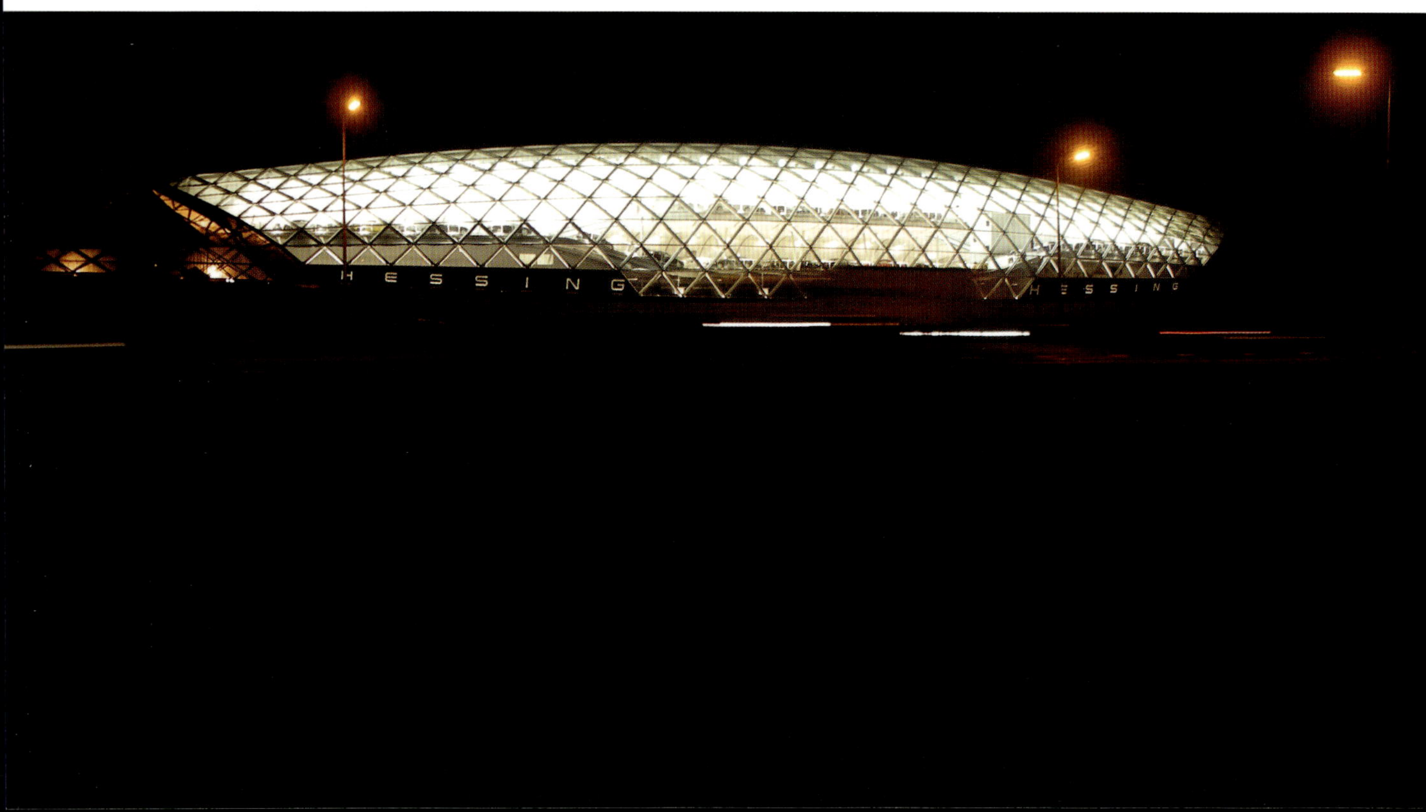

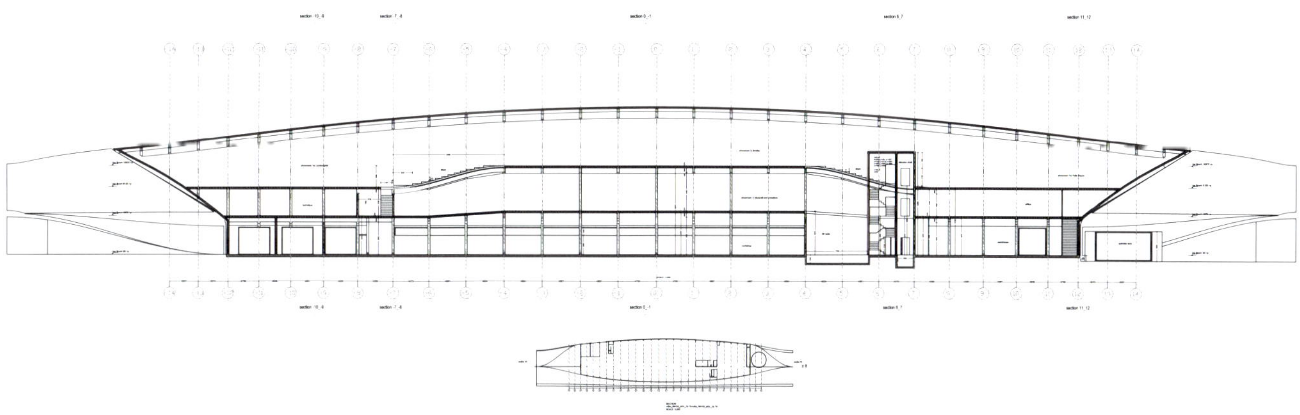

\ longitudinal section AA

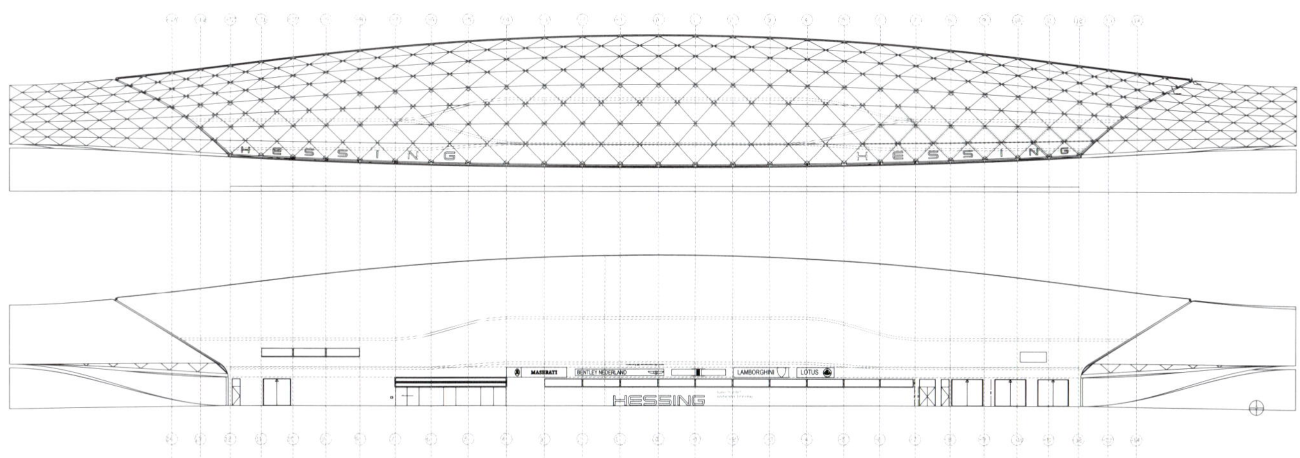

\ facade

음향 제어벽 안에 헤싱 콕핏의 혁신

건축도시계획프로젝트인 '음향 제어벽 안에 헤싱 콕핏(Hessing Cockpit in Acoustic Barrier)'은 여러 분야에서 혁신적인 내용을 담고 있다.

도시계획의 혁신

설계안은 건물의 볼륨을 방음벽 안으로 통합한다. ONL은 SPVE (도시계획설계규정)에 명시된 몇 가지 설계 규칙을 명확하게 표현한다. 건물은 시속 120킬로미터로 지나가는 자동차를 고려해 설계된 것으로, 전체 건물의 볼륨은 길이와 높이의 비가 10:1인 비율에 따라 자동차가 지나가는 방향으로 뻗어나가 있다. 계획설계에서 건물의 외벽은 접이식 라인을 따라 탄력적인 천장선과 바닥선 사이로 뻗어나간다.

건축적 혁신

설계안은 비표준건축이 대규모로 실현된 명확한 사례가 된다. NSA(Non Standard Architecture)의 기본 원리는 상응하는 모든 구성요소가 원칙적으로 다르다는 것이다. 예외가 규칙이 되었다. 두 가지 유사한 요소가 있다면, 이것은 고의가 아니고 적절한 것도 아니다. 그러나 모든 예외는 엄격하게 제한된 변수설계 시스템 속에서 발생한다. "하나의 건물에 하나의 디테일"이라는 격언이 여기에 적용된다. NSA는 대량 주문생산이라는 산업생산방식에 근거를 두고 있다. 유사한 요소의 반복은 더 이상 경제적인 이익이 아니고, 그것은 반복의 미학에 대한 유효한 주장도 아니다. 동일한 요소의 반복은 더 이상 아름다움과 동일한 것이 아니다. NSA가 원리로 삼은 아름다움은 이제 수천 개의 다른 요소들을 조정하는 것으로 형성된다. 이런 새로운 기술을 손에 넣기 위해서 ONL은 특정한 속도로 움직이는 공산품의 설계과정을 연구했다. 이런 것들 중 한 예는 곡면의 표면 속으로 사라지는 접이식 라인들이다. 비표준건축은 자연스럽게 변화하는 건축이다.

설계기술의 혁신

기하공학은 ONL의 내부에서 개발되었다. ONL은 통합된 수만 개 요소의 정확한 데이터를 전적으로 책임진다. 이것을 목적으로 ONL은 프로젝트 상세 설명서의 작성을 통해 효과적인 파일에서 공장까지(File to Factory)의 프로세스를 프로그램화했다. 설명서는 기하학적인 설계를 정확하게 묘사하고, 데이터는 데이터베이스에 저장된다. 제작기계를 조작하는 소프트웨어는 이 정보를 읽는다. 건축가의 3차원 모형과 제조업체의 생산기계는 직접 연결된다. 철골 생산업체는 이어서 절점에 대한 더 자세한 설명서를 작성했다.

건축기술의 혁신

ONL과 메이어스 스탈보우(Meijers Staalbouw)는 절점의 온점과 연결보로 이루어진 시스템을 개발했다. 각각의 상이한 공사조건으로 인해 두 개의 서로 다른 변형이 방음벽과 콕핏(Cockpit)을 위해 개발되었다. 방음벽에는 소음 누출이 없는 10kg/m^{2}인 덩어리로 이루어져야 한다는 하나의 요구조건 밖에 없다. 콕핏에는 단열유리가 설치되었고 건물의 볼륨이 부풀어오름에 따라 연결보와 삼각형 유리판의 치수는 두 배로 늘어난다. 모든 공사에서 공통적인 것은 연결보가 최대한 단순해지도록 그 복잡성은 절점 속으로 완전히 통합된다는 것이다. 삼각형 유리판 역시 평평하게 제작된다. 이중 곡면유리는 비쌀 뿐만 아니라, 전반적인 이미지에도 도움이 되지 않는다.

건축공사비의 혁신

메이어스 스탈보우는 ONL과의 협력 작업 속에서 전체 방화벽을 고정가 제품으로 공급했다. 노동력은 더 추가되지 않았다. 콕핏에 대해서도 메이어스 스탈보우는 뼈대 전체(철골공사 + 콘크리트 바닥 슬라브 + 유리 파사드 + 알루미늄 파사드)를 언급할 만한 추가 작업에 대한 요구 없이 고정가로 제공했다. 공사와 기하학이 건축에 통합된 부분이 되는 경우에 전통적인 입찰방식은 소용이 없다. 제품생산에 대한 지식, 그리고 건축가와 제조업자 간의 의사소통방식이 설계의 초기단계에서 검토될 필요가 있다. 이 단계는 입찰 때까지 연기될 수 없다. 헤싱 콕핏 건물과 방음벽은 전통적인 입찰 방식으로 시장에 나왔다면 두 배 더 비싼 건물이 되었을 것이다. 콕핏의 뼈대와 방음벽 전체는 합의한 예산 내에서 실현되었다.

Innovative in Hessing Cockpit in Acoustic Barrier

The architectural-urban planning project Hessing Cockpit in Acoustic Barrier is innnovative in several seperate fields:

Urban planning innovation

The design integrates a building volume in an acoustic barrier. ONL has defined a few clear design rules, which are specified in the SPVE(urban planning design requirements). The building is designed with the speed of a car passing at 120km/h. The integrated building volume is stretched in the direction of the flow of cars according to the formula: length: height = 10:1. In the preliminary design a surface is stretched between the elastic top and bottom lines and following the folding lines.

Architectural innovation

The design is a pure example of Non Standard Architecture realized on a big scale. Basic principle for the NSA is that all compiled components are in principle different. The exception is the rule. If there are two similar components, it is not on purpose and not relevant. But all exceptions take place in a rigidly defined parametric design system. The adagium "one building, one detail" applies here. NSA is based on the new industrial production method of mass customization. Repetition of similar elements is no longer an economic advantage and it is not a valid argument for the aesthetics of the repetition. Repetition of the same elements is not anymore identical to beauty. The beauty of the NSA principle is now in the shaping of the control of series of thousands of different elements. To acquire these new techniques ONL learned from the designing of industrial products which moves or are moved with a certain speed. One example of this are the folding lines which fade out into the curved surface. Non Standard Architecture is the architecture of smooth transitions.

Technical design innovations

The engineering of the geometry was developed in-house at ONL. ONL is entirely responsible for the precise data for the tens of thousands of integrated elements. For this purpose ONL has programmed an effective File to Factory process through the writing of project specific scripts. The scripts describe the geometry exactly and the data is registered in a database. This data is read by the software controlling the production machinery. There is a direct link between the 3d model of the architect and the production machines of the manufacturer. The steel manufacturer has subsequently written scripts for the further specification and production of the nodes.

Technical construction innovation

ONL and Meijers Staalbouw have developped a system of a point cloud of nodes and connection beams. Two different variants are developed for the Acoustic Barrier and the Cockpit, due to the different conditions for both constructions. The Acoustic Barrier has as only demand to consist of 10 kg/m2 mass without noise leak. The Cockpit has insulated glazing and through the 'inflating' of the volume the dimensions for the connection beams and the triangular glasplates are twice as big. Both constructions have in common that the complexity is completely integrated in the nodes so that the connection beams are maximally simple. The triangles are also produced as flat surfaces. Double curved glass are not only more expensive, but wouldn't enforce the overall image.

Building costs innovation

Meijers Staalbouw has offered the whole Acoustic Barrier as a product with a fixed price in close collaboration with ONL. There has been no extra labour. Also for the Cockpit Meijers Staalbouw offered the whole shell(steelconstruction + concrete floors + glass facades + aluminium facades) for a fixed price without mentionable extra work required. The traditional tendering doesn't work with designs in which construction and geometry are integral part of the architecture. The knowledge of the production and the manner of communication between the architect and the manufacturer needs to be processed in the preliminary stage of the design. This cannot be postponed till the tendering. The Hessing Cockpit building and the Acoustic Barrier would be twice as expensive if they would be tendered on the market as a traditional bidding. The shell of the Cockpit as well as the whole Acoustic Barrier are realized within the budget agreed upon.

ONL

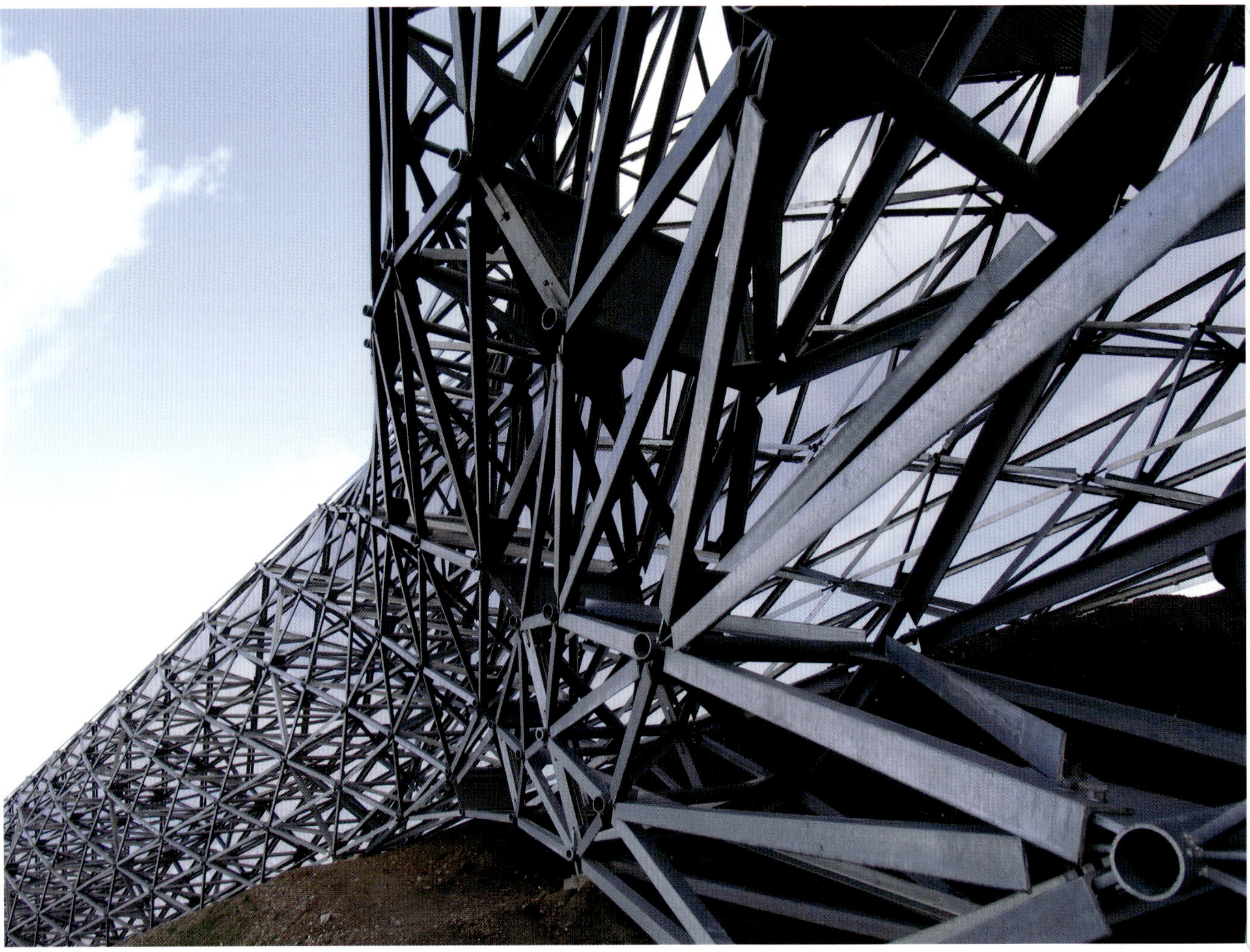

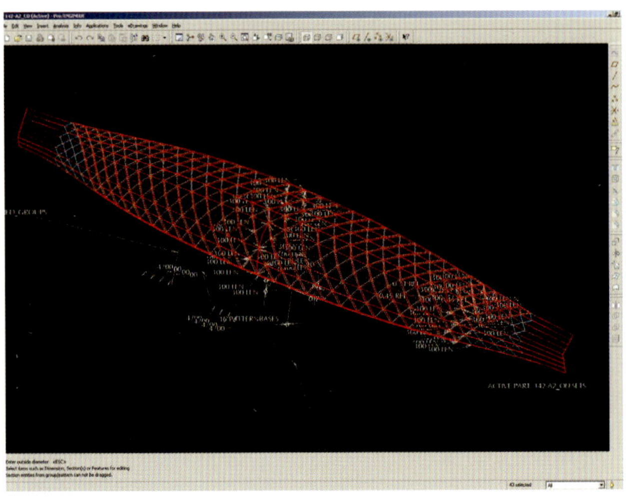

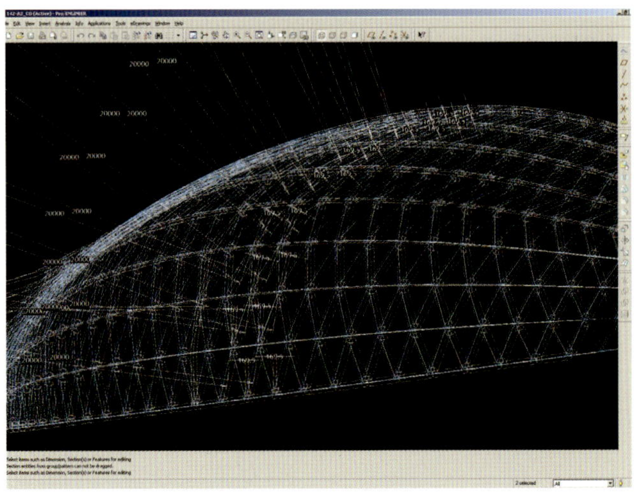

Pro/Engineer screenshots of the Cockpit
Copyright information:ONL [Oosterhuis_Lénárd]
Showing the File-to-Factory Process

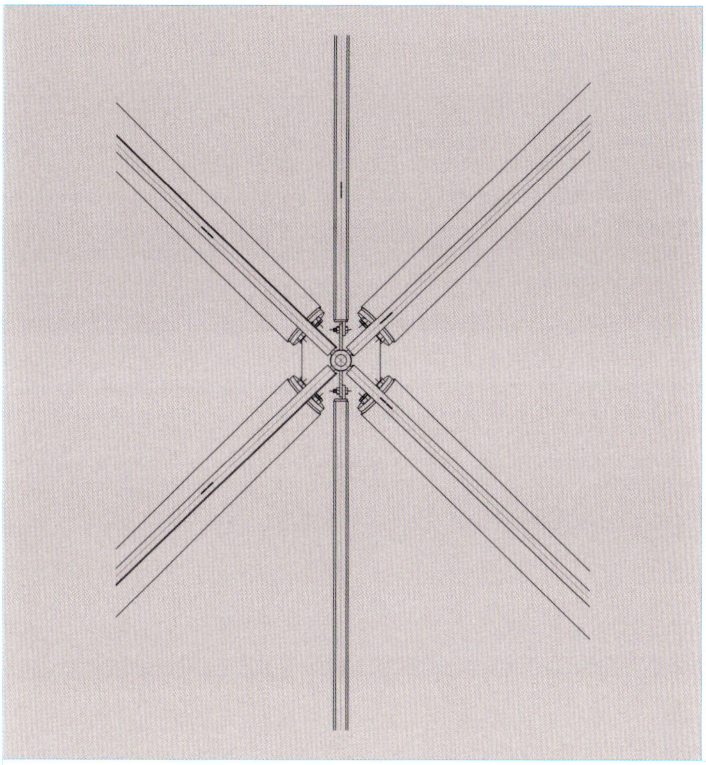

solid-node detail

node detail section

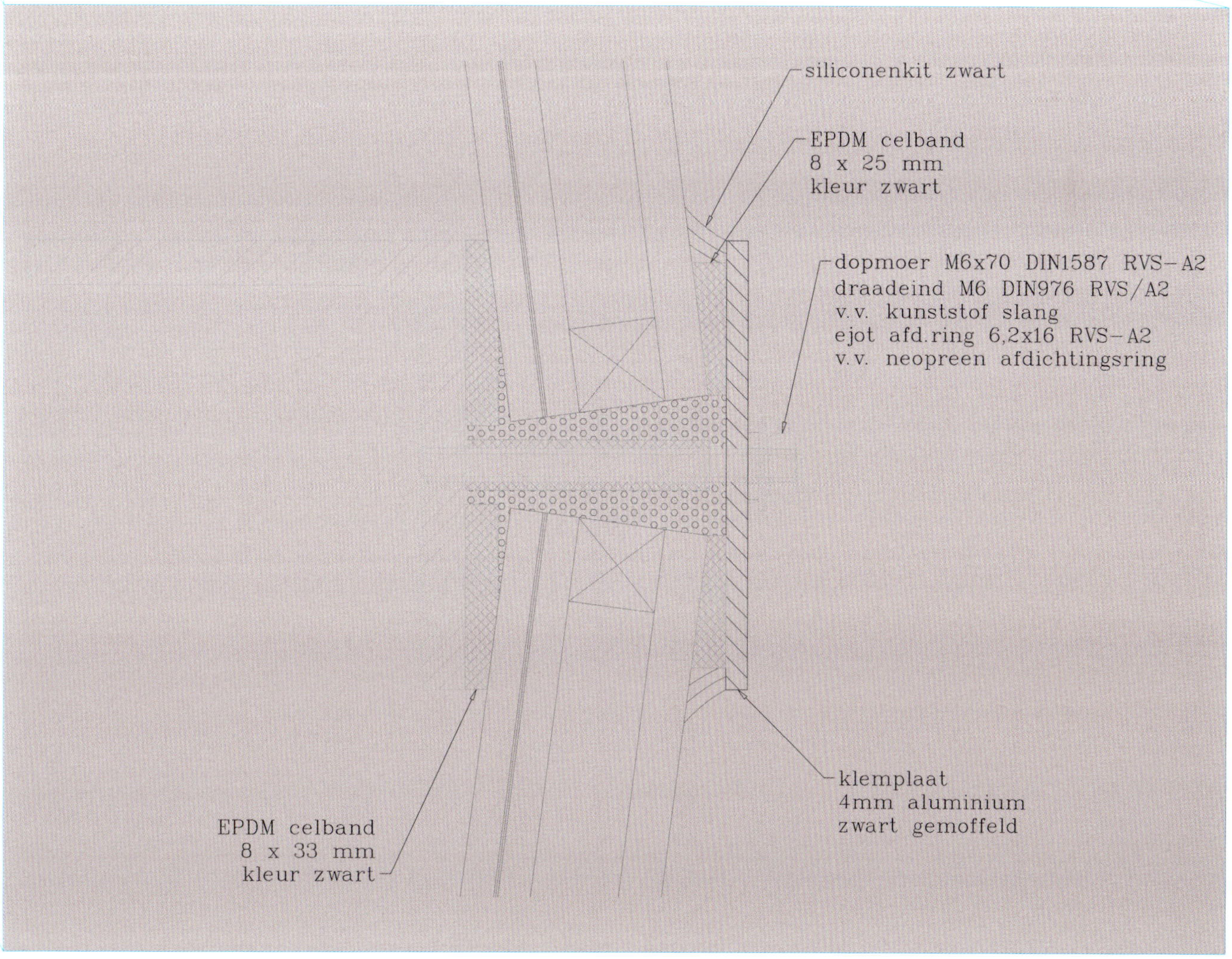

MONOLITHIC STRUCTURES

대지의 제한된 조건으로 선택된 일체식 구조의 골격을 이루는 건축물은 단지, 요구된 기능과 형태의 결합을 보여주는 것이 아니라, 외피상에 표현된 개념의 상징과 공간의 유연한 구성의 실현 등 적극적인 골조의 강조를 통한 다양한 공감각적 체험을 생성해 내고 있다. 공간을 짜는 방식에 있어서 이루어지는 건축가들의 다양한 시도와 연구는 건물의 기술과 건축의 예술 사이에서 대화 통로를 열어주고 있으며, 구조적 파사드가 놓이는 장소에서는 그것이 하나의 오브제로서 지역에 새로운 예술적 영감을 스스로 발산하고 있다.

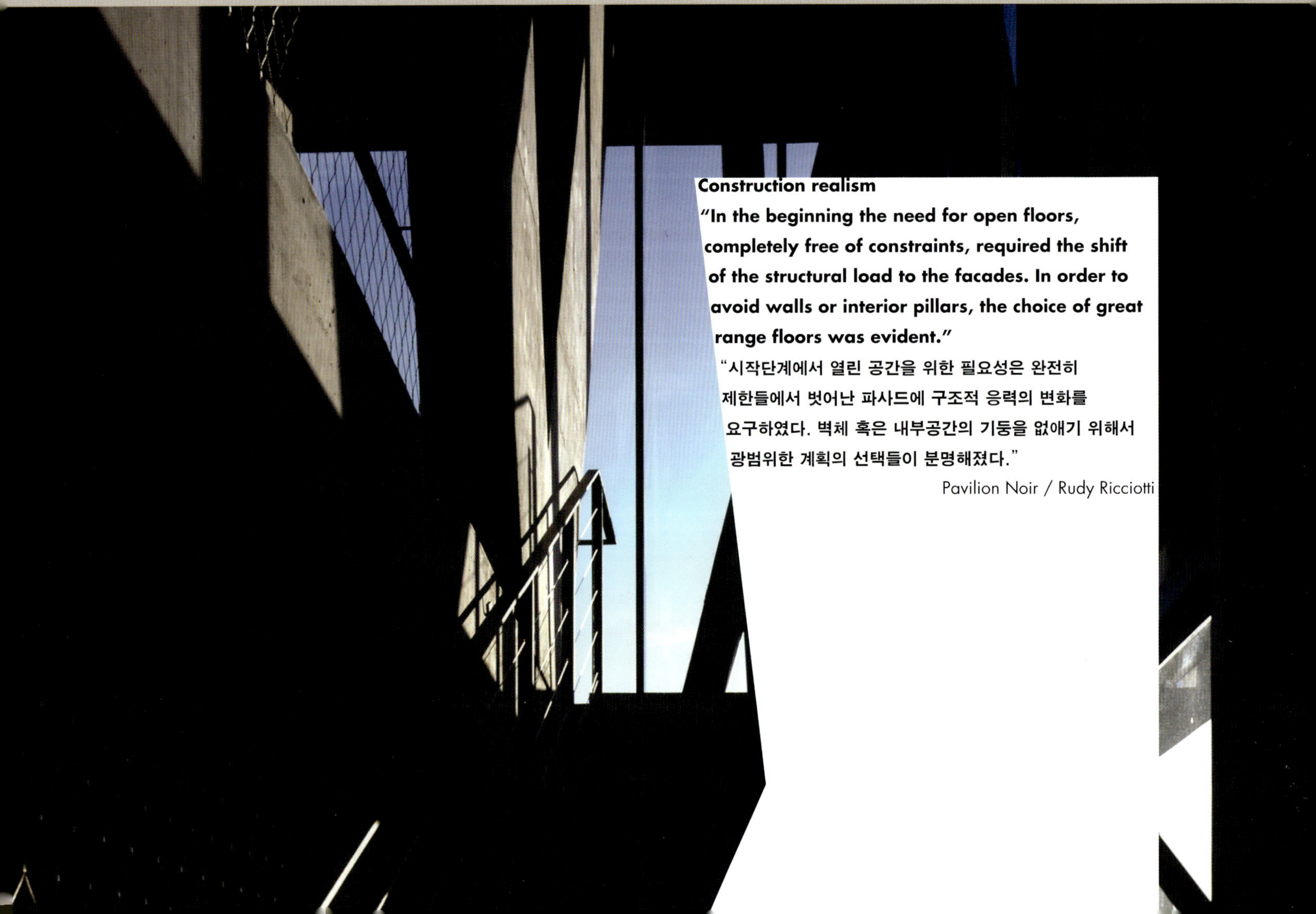

Construction realism

"In the beginning the need for open floors, completely free of constraints, required the shift of the structural load to the facades. In order to avoid walls or interior pillars, the choice of great range floors was evident."

"시작단계에서 열린 공간을 위한 필요성은 완전히 제한들에서 벗어난 파사드에 구조적 응력의 변화를 요구하였다. 벽체 혹은 내부공간의 기둥을 없애기 위해서 광범위한 계획의 선택들이 분명해졌다."

Pavilion Noir / Rudy Ricciotti

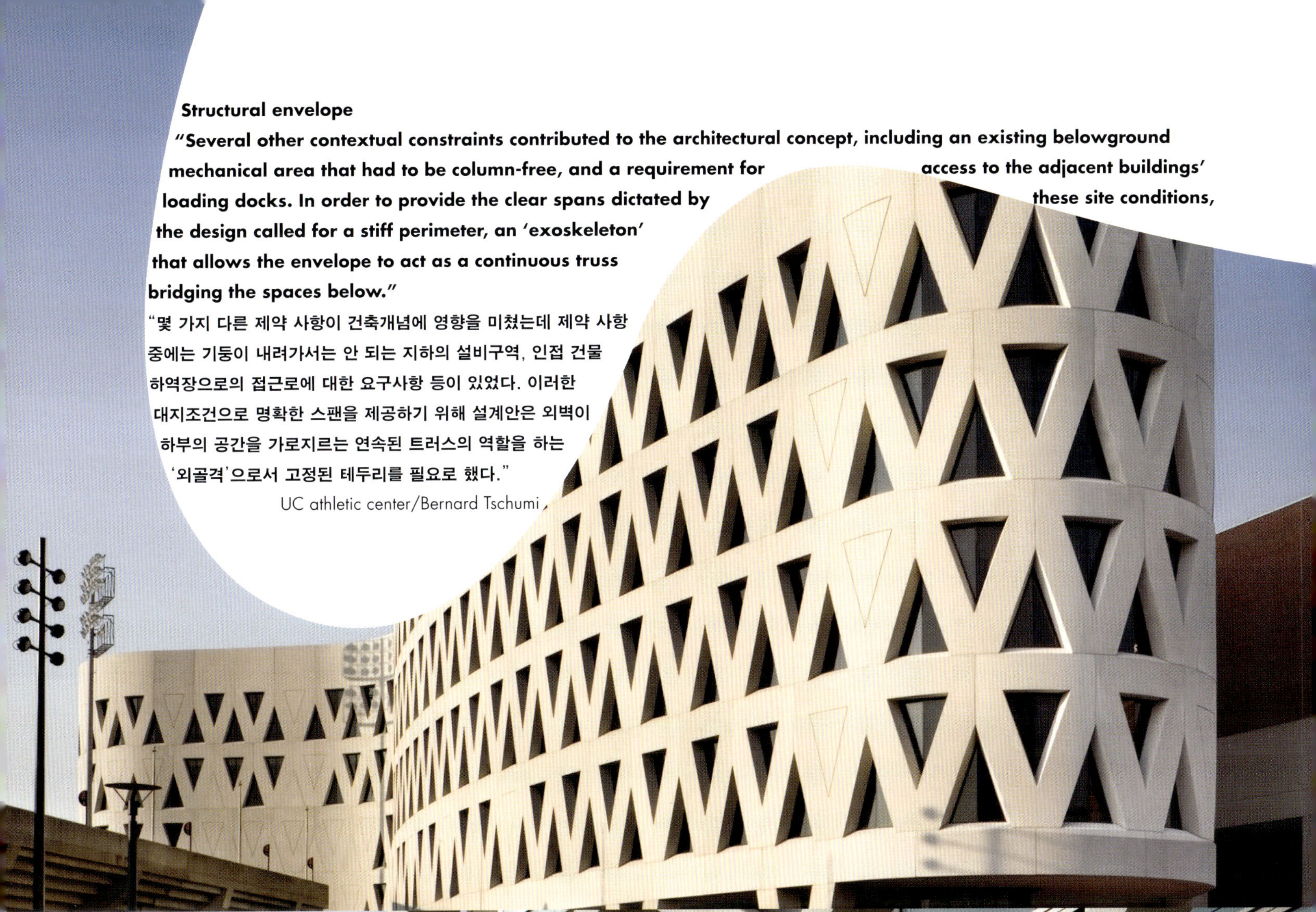

Structural envelope

"Several other contextual constraints contributed to the architectural concept, including an existing belowground mechanical area that had to be column-free, and a requirement for access to the adjacent buildings' loading docks. In order to provide the clear spans dictated by these site conditions, the design called for a stiff perimeter, an 'exoskeleton' that allows the envelope to act as a continuous truss bridging the spaces below."

"몇 가지 다른 제약 사항이 건축개념에 영향을 미쳤는데 제약 사항 중에는 기둥이 내려가서는 안 되는 지하의 설비구역, 인접 건물 하역장으로의 접근로에 대한 요구사항 등이 있었다. 이러한 대지조건으로 명확한 스팬을 제공하기 위해 설계안은 외벽이 하부의 공간을 가로지르는 연속된 트러스의 역할을 하는 '외골격'으로서 고정된 테두리를 필요로 했다."

UC athletic center/Bernard Tschumi

Iconic concrete grid

"Rafael Viñoly Architects' design addresses the paramount question of functionality, while providing the means to bridge the traditional divide between research and education. The firm's response to the urban plan which required a monolithic structure resulted in a bold cube-shaped building of striking sculptural quality, achieved by very restrained means. The building's iconic concrete grid is part of the primary structure."

"라파엘 비뇰리(Rafael Viñoly) 건축사무소의 디자인은 연구와 교육사이의 전통적 경계를 연결하는 수단을 제공하는 반면 기능성이라 대전제에 대한 의문을 언급한다. 조각품에 가까운 수준의 두꺼운 육면체형태를 만들어내는 완전히 단일한 구조체를 요구하는 도시 계획에 대하여 건축가의 회답은 아주 제한된 방법에 의해서만 이루어졌다. 건축물의 상징적인 '콘크리트 격자'는 주요한 구조물 일부이다."

Atlas Building/Rafael Viñoly

Aix-en-Provence, France

1999-2006

Architecte Rudy Ricciotti

Function_ Dance theater hall
Total floor area_ 3500m² shob
Photographer_ Architecte Rudy Ricciotti
Client_ SEMEPA- City of Aix-en Provence
Project architect_ Tilman Reichert, Raphaelle Segond
Structural Engineer_ SEV Ingenierie(Serge Voline), Aix-en Provence(Serge Voline)

PAVILION NOIR

2006년 프랑스 엑상프로방스(Aixen Provence)지역에 문을 연 파빌리온 누아르(Pavilion Noir)는 네 개의 녹음 스튜디오와 퍼포먼스 홀, 연습실로 구성된 문화시설이다. 프랑스 건축가 루디 리시오띠(Rudy Ricciotti)에 의해 설계된 이 건물은 아무런 포장 없이 거칠고 대담한 검은 구조물로 엮여져있으며, 가공되지 않은 콘크리트 구조물 사이로 인간의 움직임과 도시의 역동성이 서로 반응하고 있다.

CCN의 건설을 위해 이용 가능한 구획의 규모는 제한적이고, 어떤 이는 한정적이라고 말할 수 있을 정도이다. 이것은 일종의 도전일 뿐 아니라 기회이기도 하다.

이 프로젝트는 제한된 요구들로부터 탄생하기 때문에, 오직 표피와 뼈대로만 구성될 것이다.

건축적 경쟁(네오모더니즘적 구원자, 타락한 포스트모더니즘, 하이테크적 식민주의, 스코틀랜드식 건축적 음극)의 정신분열적인 공연에서 그것은 마치 유토피아로 혼란에 빠진 채 끝마치게 될 응급행위로 남아 있으며, 재구성하는 몸짓의 행위로 남을 것이다.

그것은 아직 가치로서의 작업과 노력의 아이디어에 동의할 필요가 있다고 한다. 건축은 그것에 대한 흥미를 잃었다.

작업이 미학적으로 부정확해지고 노력이 정치적을 타당하지 않을 것으로 여겨지고 있는 동안. 유죄와 비양심의 공간 안에 난민들, 빅토르 휴고(Victor Hugo)의 장발장(Jean Val Jean)과 같은 사람과 더 가까운 건축가들은 가치상에 최하위에 있게 된다.

CCN이라는 전쟁에서, 더 정신적인 것을 위한 연구는 오직 명백함으로부터 올 수 있었다.

똑, 똑, 누구세요? 육체 혹은 영혼…우스운 놈 혹은 관료 양반…구조 혹은 표피…최소 혹은 최대…협력 혹은 반대?

시작단계에서 열린 공간을 위한 필요성은, 완전히 제한들에서 벗어난, 파사드에 구조적 응력의 변화를 요구하였다. 벽체 혹은 내부공간의 기둥을 없애기 위해서 광범위한 계획의 선택들이 분명해졌다.

수직적 그리고 비스듬한 구조적인 강제를 관리하는 것은 유명한 지진의 법규 PS 92에 의해 필요로 하였고, 수학적인 구조의 모델링과 같은 새로운 실험을 요구하였다. 우리는 컴퓨터에 물었다. 그가 그것에 대해 생각하고 있는 것은 무엇이지?

- 건설적 리얼리즘(구조적 부담에 대한 강하정도)
- 네오-리얼리즘(실제 지진)
- 합리주의(바람의 영향), 등등…

새로운 합리주의는 건설적으로 구체화되지 않았다.

이 프로젝트의 포부는 어떠한 것도 넘치거나 모자라지도 않는, 수학의 독재에 대한 실제적 사실들의 측면에서 서 있는 것이다. 구조적 해결책들은 노력과 작업, 표피와 뼈대, 강함에 맞선 약함에 대해 공표하고 있다. 그보다 더, 우리는 할 수가 없다.

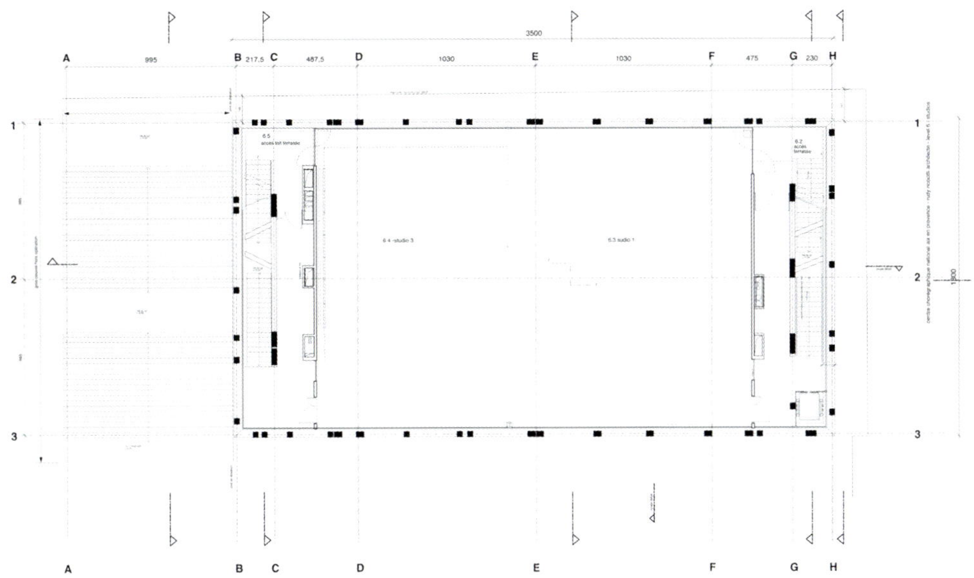

\ typical floor plan

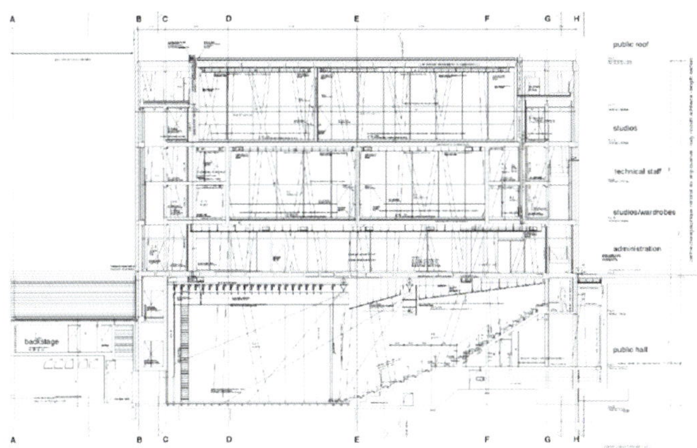

section /

The dimensions of the parcel available for the construction of the CCN are limited, one could say restricted. This represents not only a challenge, but also an opportunity.

The project is born out of the necessity to restrict; it will be skin and the bones only.

In the schizophrenic concert of the architectural quarrels(neo-modernism fireman, degenerated postmodernism, high-tech colonialist, mac-architecture-cathodic) it remains an emergency act,

which will end up being confused with Utopia, the act of reconstituting gesture.

For that it is still necessary to agree on the idea of effort and work as virtue. Architecture has lost the taste for that.

While work has became aesthetically incorrect and effort has been considered politically incorrect. Refugee in the space of culpability and bad conscience, much like Victor Hugo's Jean Val Jean, the architect keeps the foot on the coin.

In this battle of CCN, the search for more soul can only come from the obvious.

Knock, knock. Who is there? Body or spirit...fish or bureaucrat...structure or envelope... minimal or minimum... collaboration or resistance?

In the beginning the need for open floors, completely free of constraints, required the shift of the structural load to the facades. In order to avoid walls or interior pillars, the choice of great range floors was evident.

The management of the vertical and oblique structural constraints necessitated by the famous seismic regulation PS 92, required a new exercise - the mathematical modeling of the structure. We asked the Computer. What does he think about it?

The answer was prompt:

- constructive realism(descent of the structural charges),
- neo-realism(virtual earthquake),
- rationalism(efforts of the wind), etc...

A new rationalism constructively unspecified.

The ambition of this project is to stand in the face of the objective facts of the dictatorship of mathematics, nothing more and nothing less. The structural solutions celebrate effort and work, skin and bones, weak against strong. More than that, we cannot do.

Rudy Ricciotti

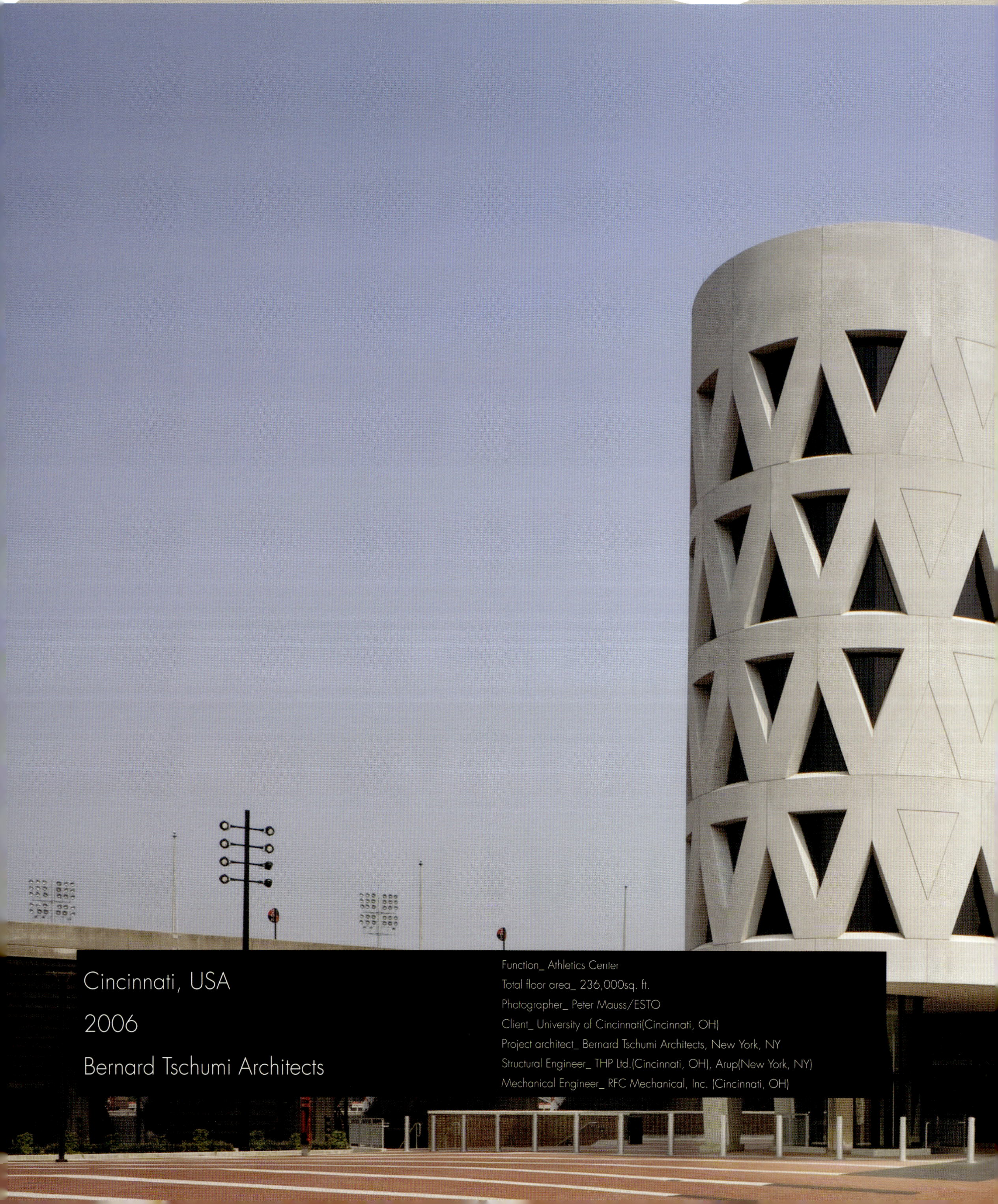

Cincinnati, USA

2006

Bernard Tschumi Architects

Function_ Athletics Center

Total floor area_ 236,000sq. ft.

Photographer_ Peter Mauss/ESTO

Client_ University of Cincinnati(Cincinnati, OH)

Project architect_ Bernard Tschumi Architects, New York, NY

Structural Engineer_ THP Ltd.(Cincinnati, OH), Arup(New York, NY)

Mechanical Engineer_ RFC Mechanical, Inc. (Cincinnati, OH)

RICHARD E. LINDNER ATHLETICS CENTER

신시내티(Cincinnati) 대학교의 체육센터는 독립적인 삽입물, 혹은 주변 상황 속의 자유로운 형태를 가진 구조물로 묘사할 수 있다. 대학 측은 원래 체육센터가 운동장 필드 사이 어디에도 위치할 수 있어야 한다고 언급했고, 우리가 모든 것을 다시 동등하고 일반적인 한계 속에 집어넣도록 만들었다. 정말 자유롭게 서 있는 건물을 만들 수도 있었지만, 우리는 체육센터(체육과 학문적 활동의 중심으로서 의도된)를 체육관과 대형 실내 경기장 사이의 좁은 틈 속, 문자 그대로 캠퍼스의 중심에 건축하기로 결정했다. 센터의 곡면 형상은 대지의 엄격한 제약 조건을 활용하기 위해 고안되었고, 그 결과 제안된 건물과 기존 체육관, 운동 경기장, 레크레이션 센터 사이에는 역동적인 짜투리 공간들이 생겨났다.

5층 규모의 신축 체육센터는 대학 캠퍼스의 북측과 남측 출입구를 연결하고, 인접한 기존 건물들의 복잡한 배열을 통합하는 허브 역할을 담당한다. 결과적으로 체육센터의 중심에 있는 아트리움은 체육활동의 주요한 종착점이 되고, 티켓 판매소가 모여 있으며, 학생, 지도자, 관리자들의 움직임을 걸러주는 역할을 한다. 또한, 건물은 최상층에는 사무실들, 강의실, 미술관, 기념품점, 연습용 체육관, 운동치료시설, 연습실, 지하에는 라커룸 등을 포함하고 있다. 주로 운동선수들이 건물을 이용할 것이지만, 새 강의실, 교실, 학부클럽, 테니스장, 축구장은 더 광범위한 대학 구성원들이 함께 공유할 것이다.

개념의 구체화: 구조적 외피

몇 가지 다른 제약 사항이 건축개념에 영향을 미쳤는데, 제약 사항 중에는 기둥이 내려가서는 안 되는 지하의 설비구역, 인접 건물 하역장으로의 접근로에 대한 요구사항 등이 있었다. 이러한 대지조건으로 명확한 스팬을 제공하기 위해, 설계안은 외벽이 하부의 공간을 가로지르는 연속된 트러스로서의 역할을 하는 '외골격'으로서 고정된 테두리를 필요로 했다.

사선격자프레임인 '다이아그리드(diagrid)'가 개발되었는데, 이것은 기둥을 구성하는 요소들이 마름모꼴 유형으로 비교적 촘촘하게 배치된 효율이 높은 구조 시스템이다. 다이아그리드의 재료는 575개의 판넬을 사용한 프리캐스트 콘크리트 클래딩을 통해 단열 처리된 구조적 철제 프레임이다. 외피의 삼각형 모양은 주변 상황 속에서 자유로운 형태와 '체육활동과 관련' 된 건물의 특징을 더욱 강조한다.

The Athletic Center at the University of Cincinnati can be described as either a freestanding infill or a contextual free-form structure. The university originally stated that the center could be located anywhere among the sports fields, provided we put everything back within the same general confines. While a truly freestanding building was a feasible option, we decided to locate the Athletic Center(intended to be an epicenter of athletic and academic activities) literally at the heart of the campus, in a narrow gap between a stadium and a large indoor arena. Its unusual curvilinear shape is designed to take advantage of the tight constraints of the site, resulting in dynamic residual spaces between the proposed building and the existing stadium, sports fields, and recreation center.

The new five-story Athletic Center links the north and south entrances of the university campus and acts as a unifying hub for the complex array of adjacent existing buildings. Consequently, the atrium at the core of the Athletic Center becomes a central terminal for athletic events, housing a centralized ticket office and acting as a filter for students, coaches, and administrators. The building also contains offices on the top floors, an auditorium, museum, gift shop, practice gymnasium, sports medicine facilities, training rooms, and locker rooms below ground. Although the building will be used mainly by athletes, the new auditorium, classrooms, faculty club, tennis courts, and soccer fields will be shared by the greater university community.

Materialization of Concept: A Structural Envelope

Several other contextual constraints contributed to the architectural concept, including an existing belowground mechanical area that had to be column-free, and a requirement for access to the adjacent buildings' loading docks. In order to provide the clear spans dictated by these site conditions, the design called for a stiff perimeter, an 'exoskeleton' that allows the envelope to act as a continuous truss bridging the spaces below.

A diagonal grid frame, or 'diagrid,' was developed -a highly efficient structural system in which column elements are laid out in a diagonal pattern at relatively close spacing. The material for the diagrid is an insulated structural steel frame with precast concrete cladding using 575 panels. The envelope's triangulated expression re-inforces the concept of the contextual free form and the building's 'athletic' character.

Bernard Tschumi

level 700 /

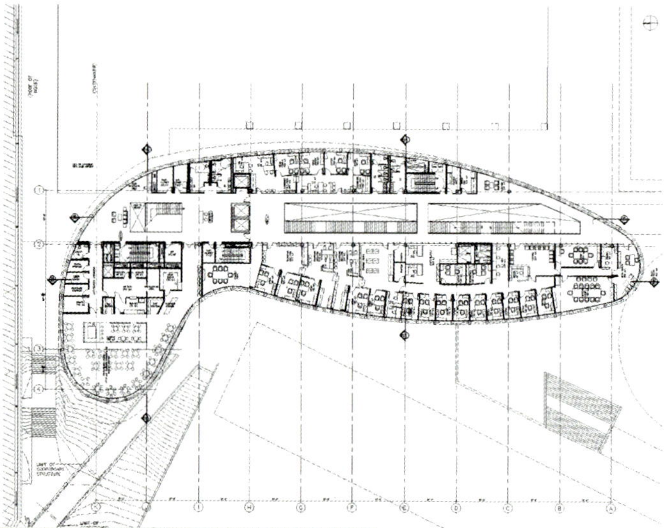

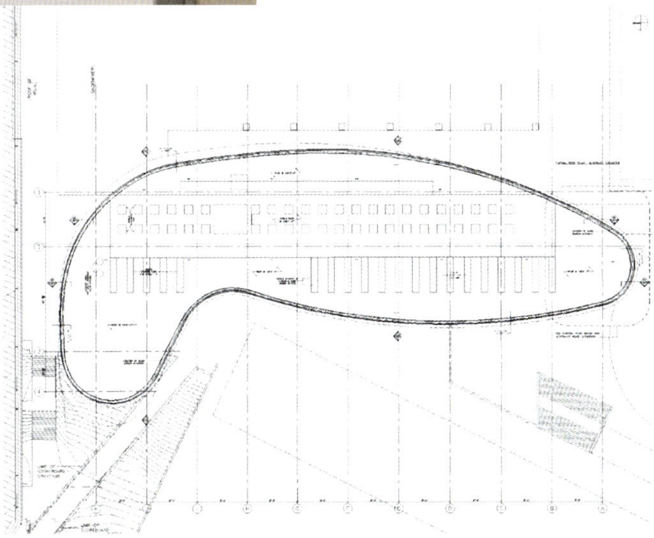

roof plan /

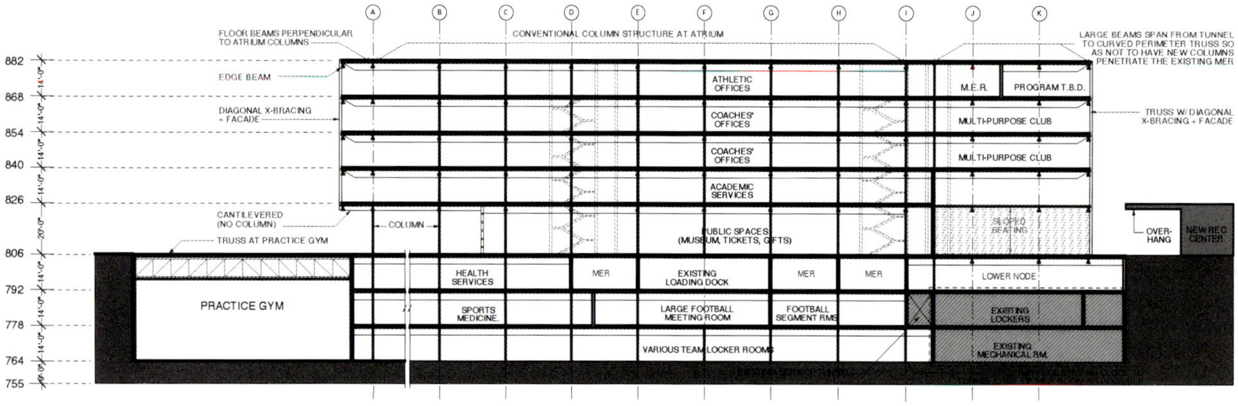

\ longitudinal section, distribution diagram

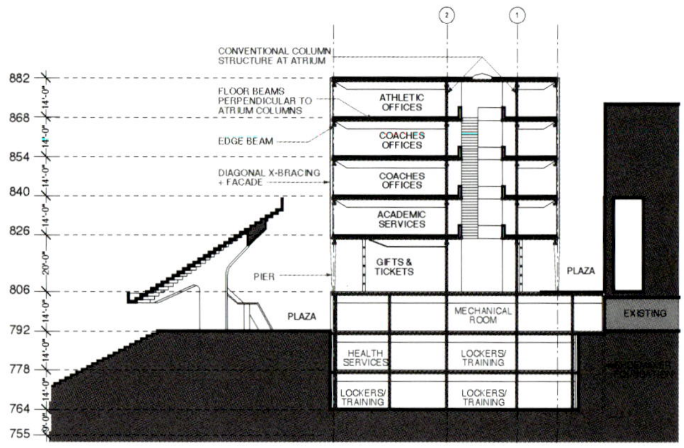

\ structural cross section thru south-athletic wing

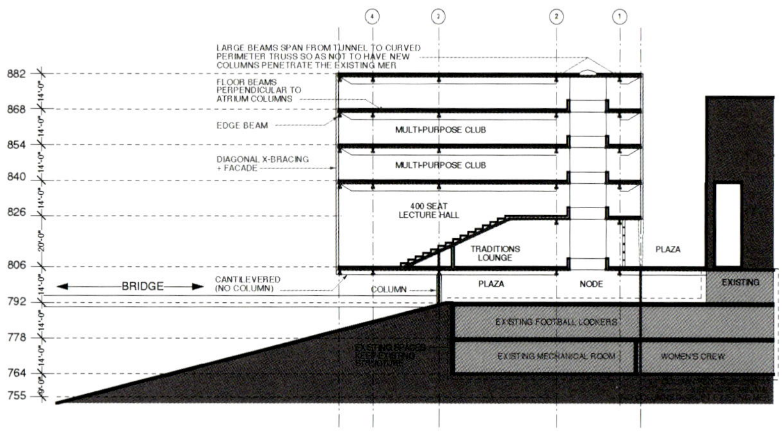

\ structural cross section thru north-athletic wing

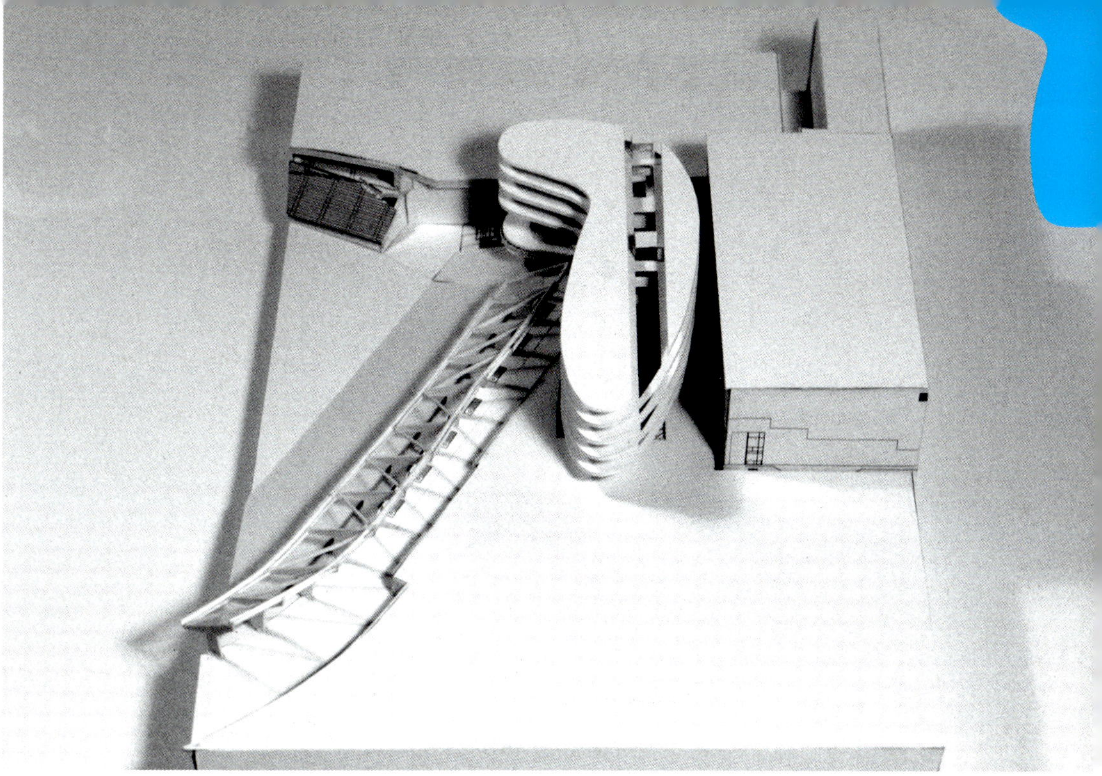

\ model

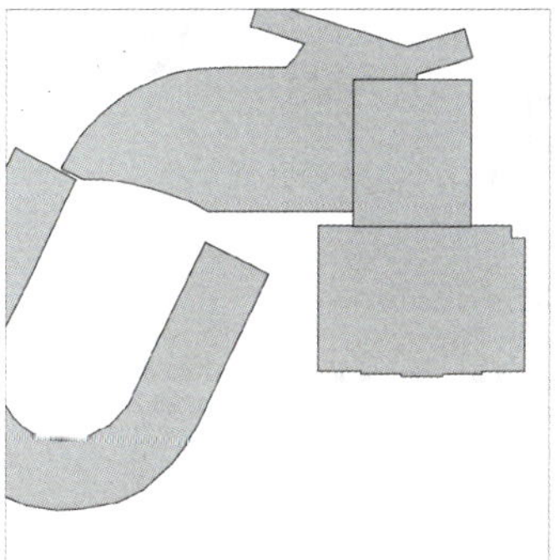

Site

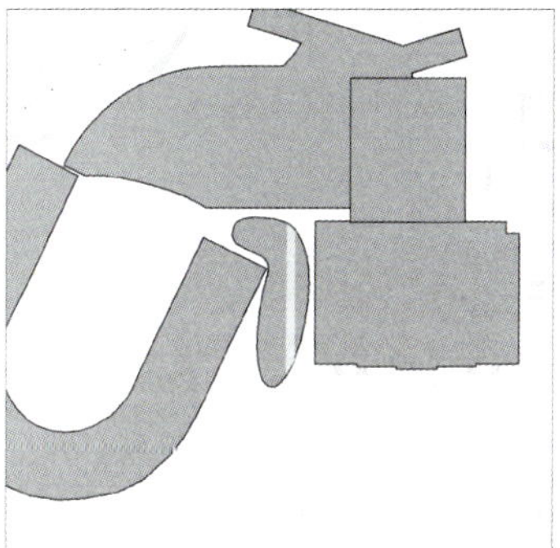

Free standing infill

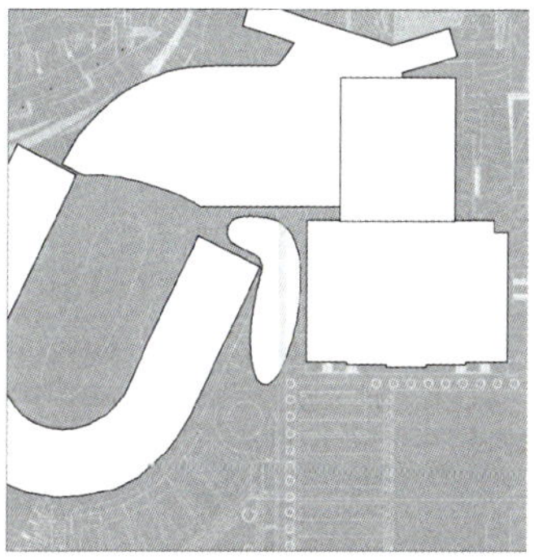

Contextual Freeform

\ site

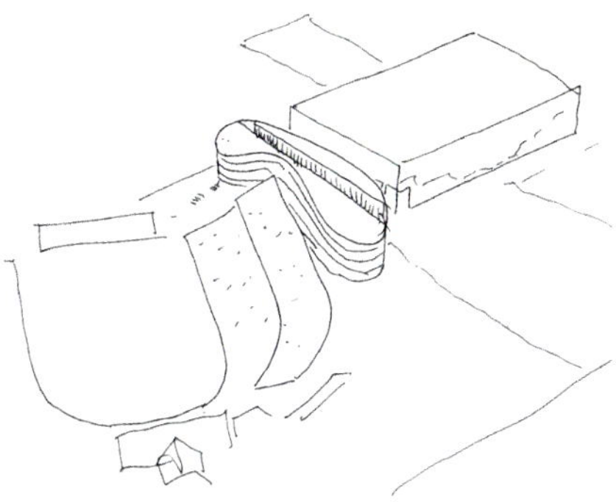

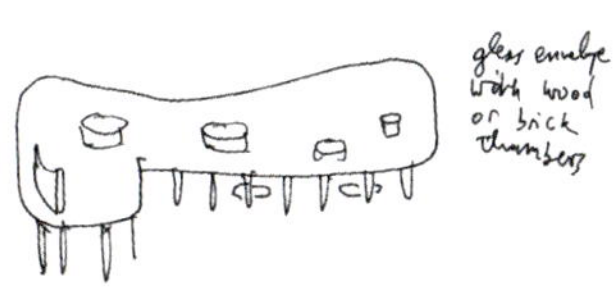

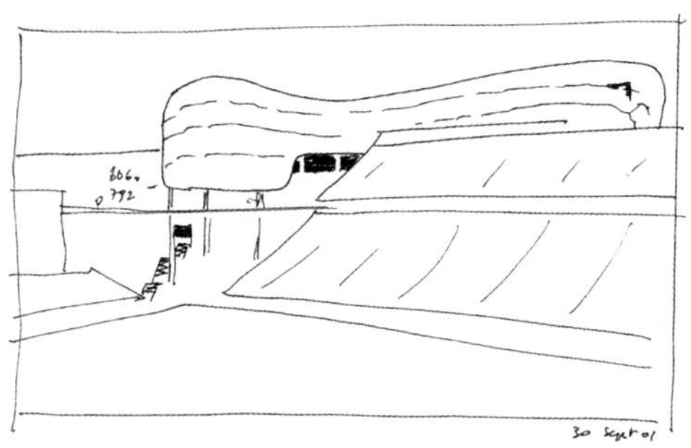

\ sketch

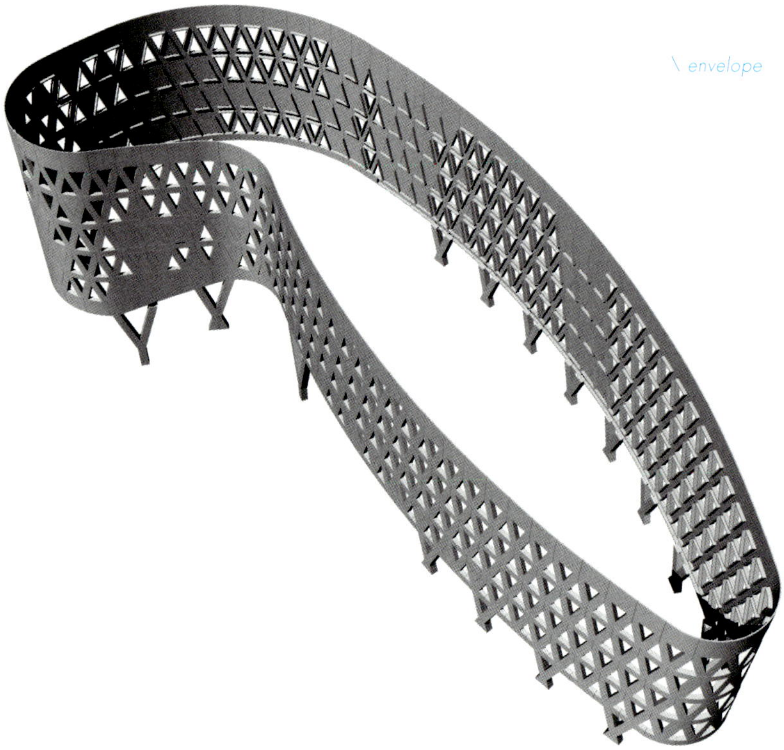

구조적 외피:
프리캐스트 콘크리트 클래딩의 구조적 다이아그리드 파사드. 구조적 파사드는 기존 설비 공간과 하역장을 지나가야 하는 필요에 의해 유추된 것이고, 체육센터에 강력하면서도 일반인들이 쉽게 인식할 수 있는 이미지를 제공하기 위한 것이다.

A Structural envelope:
Structural diagrid facade with precast concrete cladding. The structural facade is derived from the need to span over existing mechanical spaces and loading dock and to provide a strong, publicly legible image for the Athletic Center.

RICHARD E. LINDNER ATHLETICS CENTER

precast panel variation diagram

precast panel detail

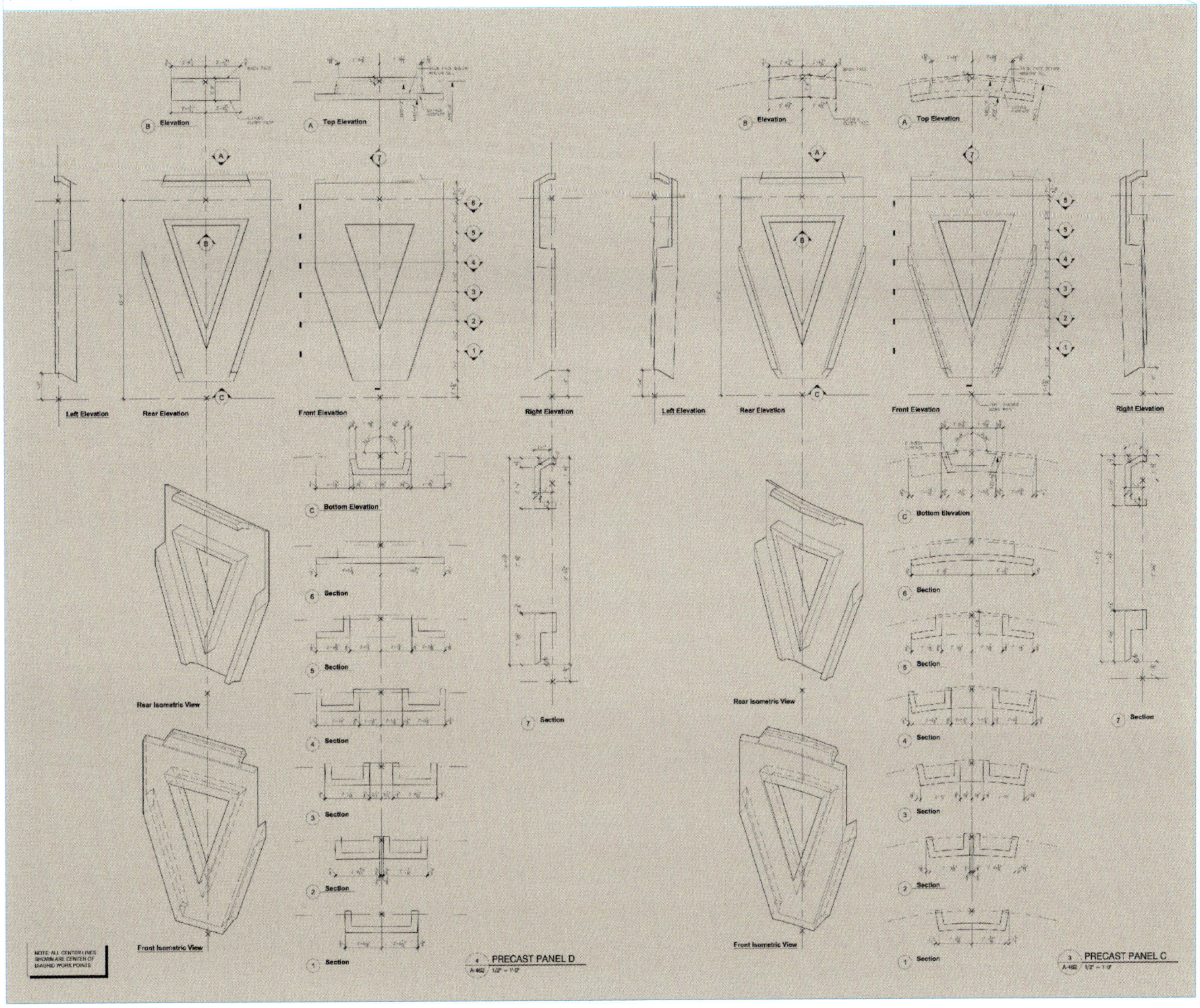

Wageningen, the Netherlands

2007

Rafael Vinõly Architects

Total floor area_ 105,000m²
Function_ Academic Laboratory
Project Architect_ Rafael Vinõly, Jay Bargmann, John Drew, Mariana Kolova
Structural Engineer_ Pieters Bouwtechniek
Photographer_ Luuk Kramer
Client_ Wageningen University and Research

ATLAS BUILDING

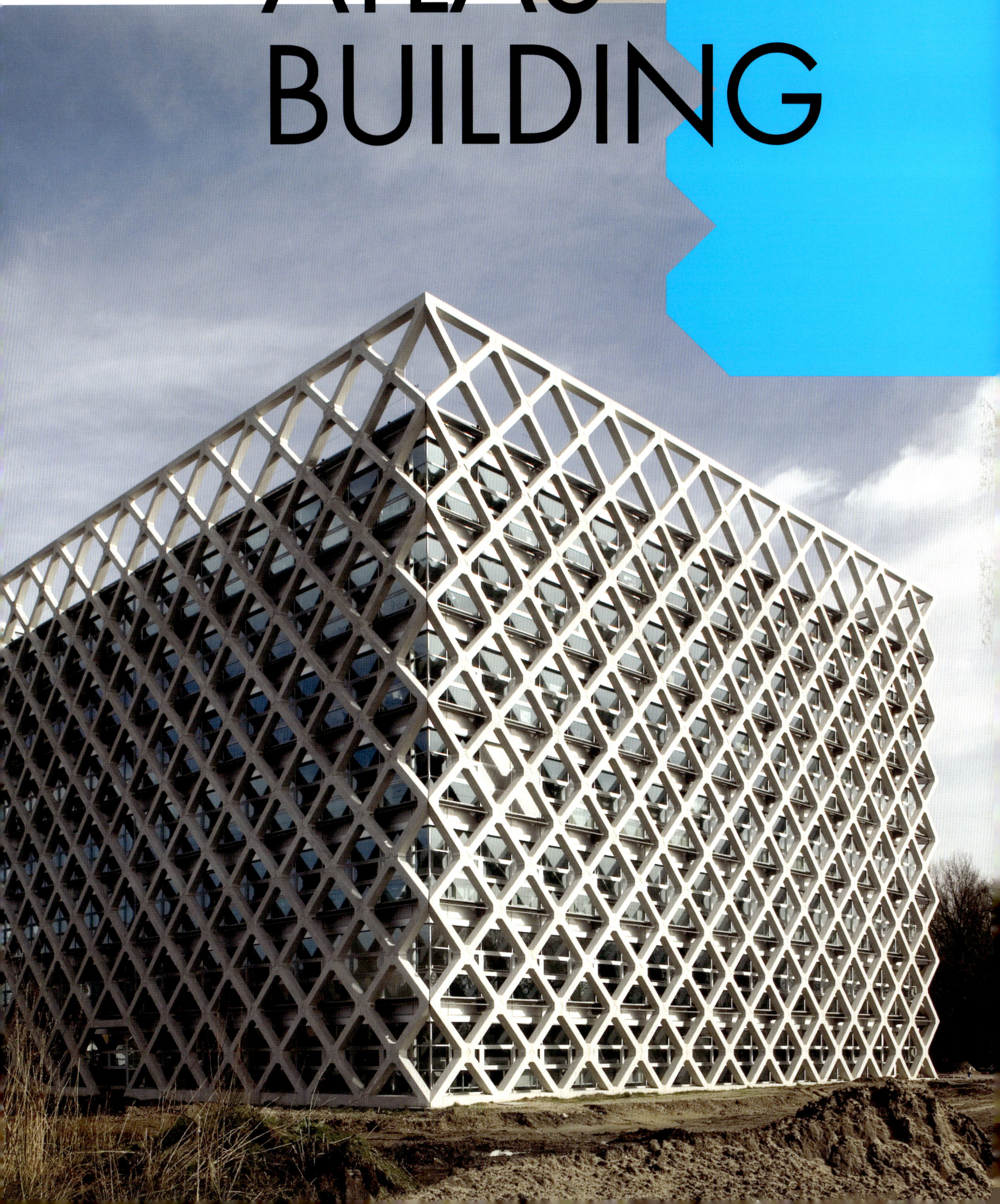

1918년에 설립된 바게닝겐대학(Wageningen University)은 환경, 동물, 식물학, 농업기술 그리고 식품과학 분야에 초점을 맞춘 세계를 선도하는 연구센터와 교육 기관으로 발전해왔다. 바게닝겐대학은 현재 대학 활동의 많은 부분을 바게닝겐지역의 북쪽에 자리한 새로운 캠퍼스, 'the Centrum de Born'으로 통합하기로 하였다.
그 지역은 현재 불모지이기 때문에 새로운 캠퍼스 개발은 라인강(Rhine) 계곡과 Hoge Velwe 공원 사이를 연결하는 녹지 통로를 제공할 수 있게 될 것이다.
새로운 캠퍼스는 대학 입구에 과학의 축 위에 위치해 있는 기존의 건물의 반대편 위치에 새로운 랜드 마크로서 연구 센터를 수용하게 된다. 신축연구센터의 3분의 1은 토양 연구실의 전용공간으로, 나머지는 사무실로 계획되어 연구와 개발을 목적으로 하는 한 단계 앞선 기술적인 플랫폼을 제공하게 될 것이다. 새로운 센터는 바게닝겐 대학의 열망을 재현하고 실현하게 될 것이며, 우리의 일상에서 항상 일어나고 있는 환경 분야의 도전적이고 중요한 과제에 대한 진취적인 발판을 제공하게 될 것이다.
라파엘 비놀리(Rafael Viñoly) 건축사무소의 디자인은 연구와 교육 사이의 전통적 경계를 연결하는 수단을 제공하는 반면 기능성이라는 대전제에 대한 의문을 언급한다. 조각품에 가까운 수준의 두꺼운 육면체형태를 만들어내는, 완전히 단일한 구조체를 요구하는 도시 계획에 대하여 건축가의 회답은 아주 제한된 방법에 의해서만 이루어졌다. 건축물의 상징적인 '콘크리트 격자'는 주요한 구조물 일부이다.
작업공간들은 연구원들과 학생들 사이를 비형식적으로 상호연결 할 수 있도록 갤러리들과 인접한 실들이 다리로 연결되면서 중앙 아트리움 주변으로 구성된다. 유리로 된 지붕은 향후 발생할 수 있는 기계적 시스템의 요구 상황과 내부온도가 과열되는 것을 방지하기 위하여 자연 채광이 북측 면에서만 들어올 수 있도록 한다. 아트리움은 다양한 활동들은 위한 넓고 유연한 공간을 제공할 것이다.

Founded in 1918, Wageningen University has evolved into one of the world's leading education and research centers, focusing on the environmental, plant, animal, agrotechnological, and food sciences. The University has chosen to consolidate much of its activities onto a new campus, the Centrum de Born, north of the town of Wageningen. The location is a barren field and its development will contribute to the green corridor connecting the Rhine Valley and Hoge Velwe Park. It is this campus that houses the new research center, which is situated opposite the existing building on the Science Axis, a landmark location at the campus' entrance. The new center will provide an advanced technological platform for research and development with one third devoted to soil research laboratories and the rest as designated office space. The center will represent - and realize - the University's aspirations and provide a progressive setting for the challenging and important environmental work that will take place there on a daily basis.

Rafael Viñoly Architects' design addresses the paramount question of functionality, while providing the means to bridge the traditional divide between research and education. The firm's response to the urban plan which required a monolithic structure resulted in a bold cube-shaped building of striking sculptural quality, achieved by very restrained means. The building's iconic concrete grid is part of the primary structure. Work spaces are organized around a central atrium with bridges connecting the galleries ond adjacent floors, encouraging informal interaction among scientists and students. A glass roof permits light to enter from the north side only, thereby avoiding overheating and the need for any further mechanical systems. It is intended that the atrium will provide a vast flexible area for multiple activities.

Rafael Viñoly

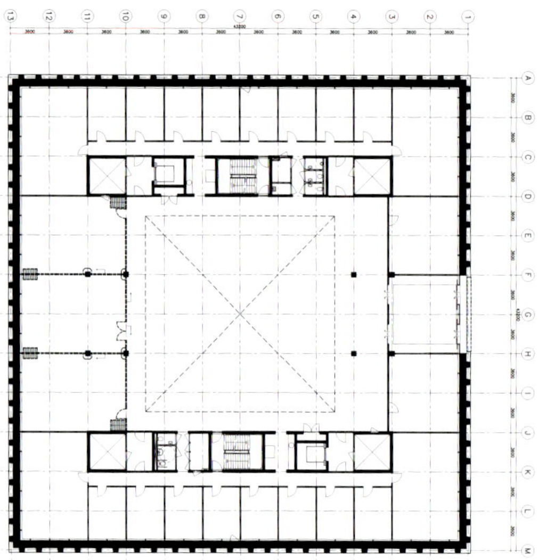

\ 1st floor plan

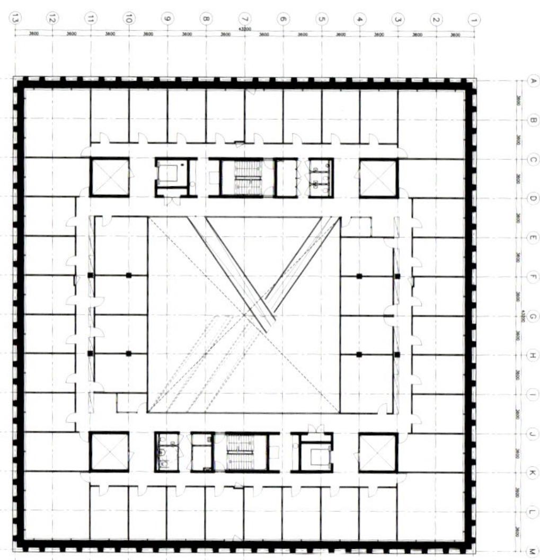

\ 4th floor plan

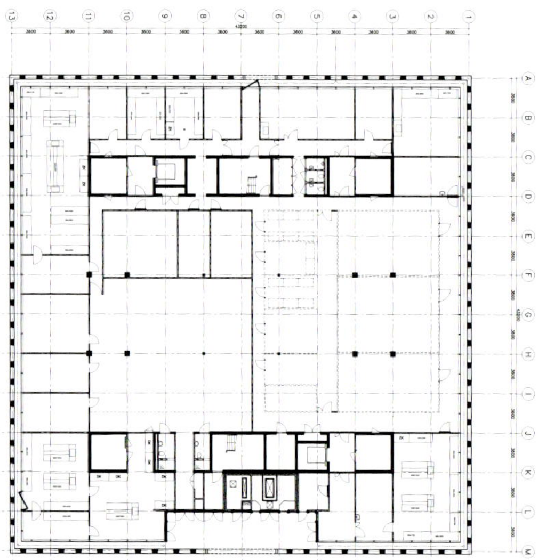

\ ground floor plan

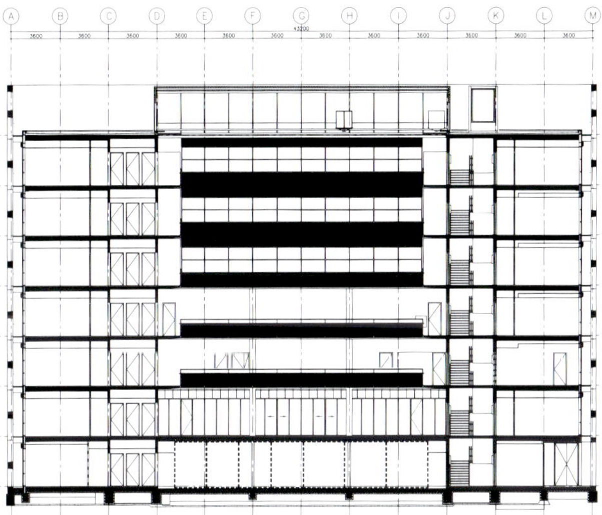

section /

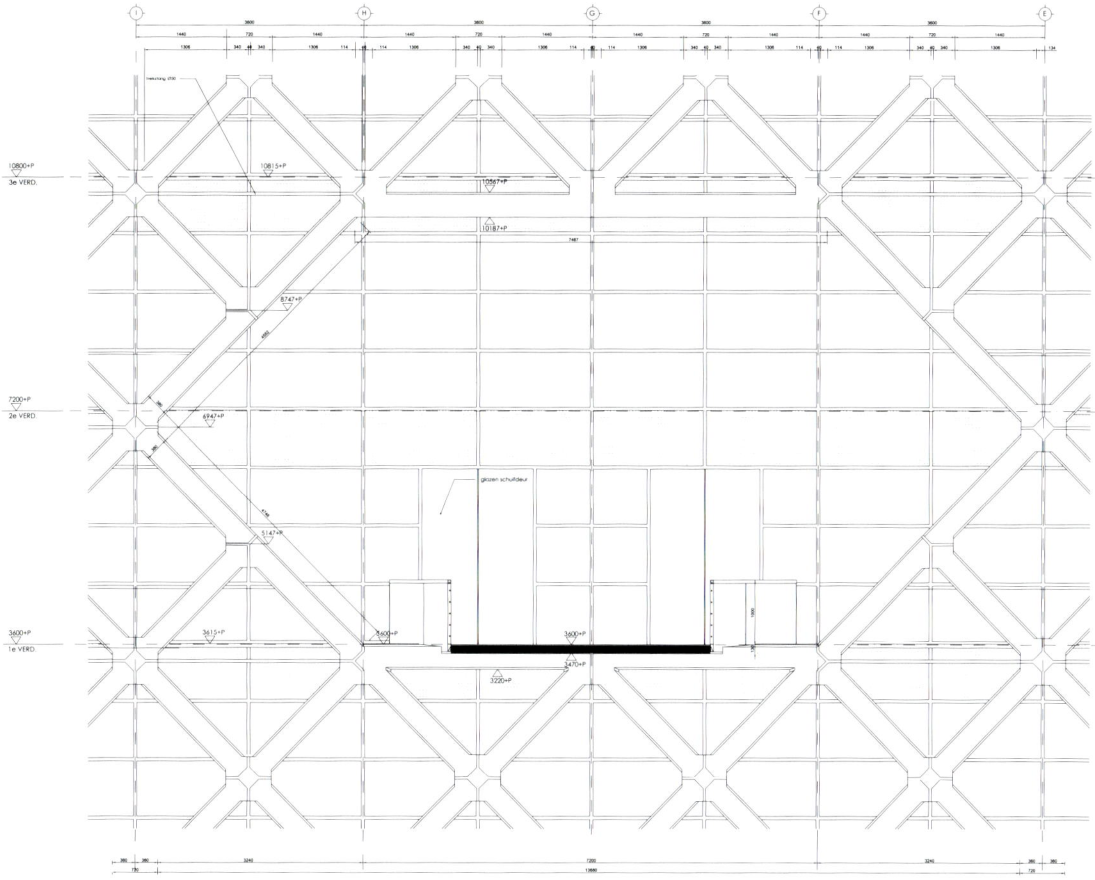

\ prefab concrete grid

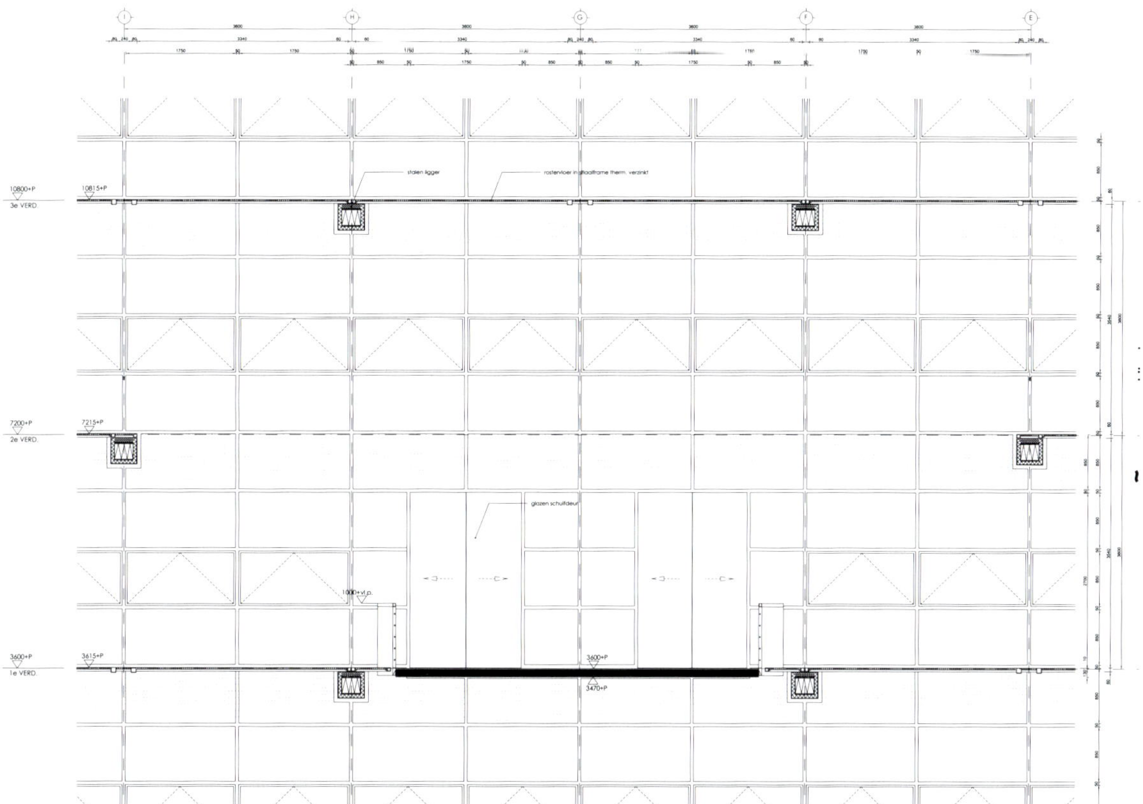

\ entry elevation

구조

아틀란스 빌딩(Atlans building)의 모서리 측면에는 8자 모양으로 생긴 10톤짜리 두 개의 다이아몬드로 구성되었다. 조밀한 콘크리트가 사용되어 철근을 가까이 배열하면서 부재의 단면크기를 400 x 380mm까지 줄일 수 있게 하였다. 프리캐스트 부재의 단면은 전면부의 오염을 방지하기 위한 형태로 만들어졌다. 상부표면으로 흐르는 배수부분은 뒤쪽으로 흐르도록 단면이 구성된 굴곡 점이 위치하는 바닥 쪽으로 물이 흘러내리도록 하면서, 밑바닥으로의 배수는 전면부의 모서리로 떨어지는 물의 양을 줄어들게 한다. 격자건물의 구조는 프리캐스트 콘크리트(Precast concrete)의 결합체 즉, 콘크리트와 강철의 결합으로 이루어진다. 외부 격자부재로부터 내부 기둥과 콘크리트 코어까지 속이 빈 강철의 보로 연결된다. 코어부분은 건물전체의 안정성을 제공하기 때문에 바깥쪽의 프리캐스트 격자로 어떠한 응력도 미치지 않게 된다. 8자 모양 부재는 보가 연결부로부터 떨어져 8자 모양 부재의 중심에 연결되도록 하기 위하여 고안되었다. 한 개의 프리캐스트 부재는 옆의 부재와 미끄러짐이 없도록 연결부와 맞닿아 위치하여 삼각형의 돌출부와 맞물리도록 한다. 8자 모양의 철근은 부재들을 하나로 만들며, 에폭시 그라우트로 적착시켰다. 프리캐스트와 내부 구조체 사이에 차이가 없는 한, 20mm의 틈은 내구성을 위한 것이었다. 이 부재들은 모두 1mm 이하의 오차를 가진 규격으로 생산되었다. 네입면은 1층 바닥에 90미터 길이의 출입구가 연결된 북측 면에 뚫린 개구부로부터 동일하게 떨어져있다. 각 입면에 실린 하중은 건축물의 바닥 면의 구석에서 그것의 정점에 이르는 삼각형으로 집중된다. 북측 면의 개구부는 프리캐스트 부재들에 과중량이 걸리는 것을 피하기 위해 삼각형 내부로 이동시켰다. 입면은 프리캐스트의 뒤틀림에 의한 외부로 밀어내는 힘이 다음 부재에 의해서 지지된다. 특히, 모서리 가까이에 상부 측 모서리에서 불균형한 전단력은 너무 커서 각각의 프리캐스트 부재는 수평의 아연도금 강철 막대부재를 바깥쪽 모서리와 함께 엮었다. 콘크리트에는 밝은 흰색의 착색제인 티타늄 산화물을 이용한 색이 입혀졌다. 착색제는 콘크리트 내에 2%의 비율로 첨가되었다. 콘크리트는 구조체는 자체적으로 내연성을 가진다. 내연성을 가진 판재의 부재에 의해서 아랫면까지 보호되는 강철보들 사이로 프리캐스트 바닥 면이 연결된다. 건물의 표면을 뚫은 곳은 열손실을 방지하기 위해서 단열재를 씌우고, 알루미늄으로 덮는다. 건물의 중심부에는 최대로 높인 북향의 채광을 받는 중정이 있다. 화재시 탈출은 두 개의 보조 통로를 통해서 두 군데 코어에 있는 계단실로 연결된다. 프리캐스트 부재들의 연결과 조립된 부재들 사이에 복잡한 응력들이 결합되기 때문에 부재들은 매우 빠르게 조립된다.

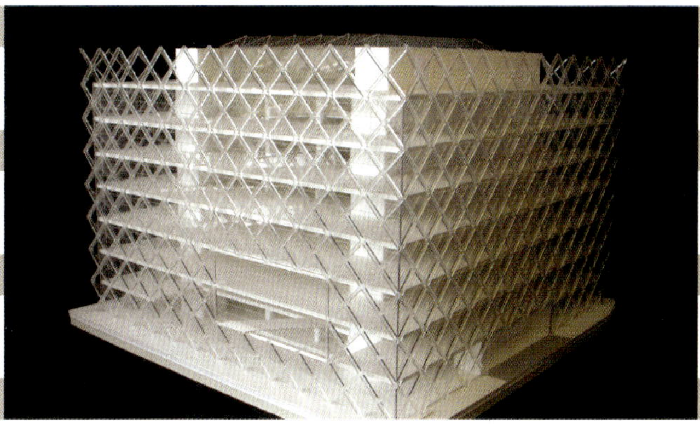

wag detail 1a

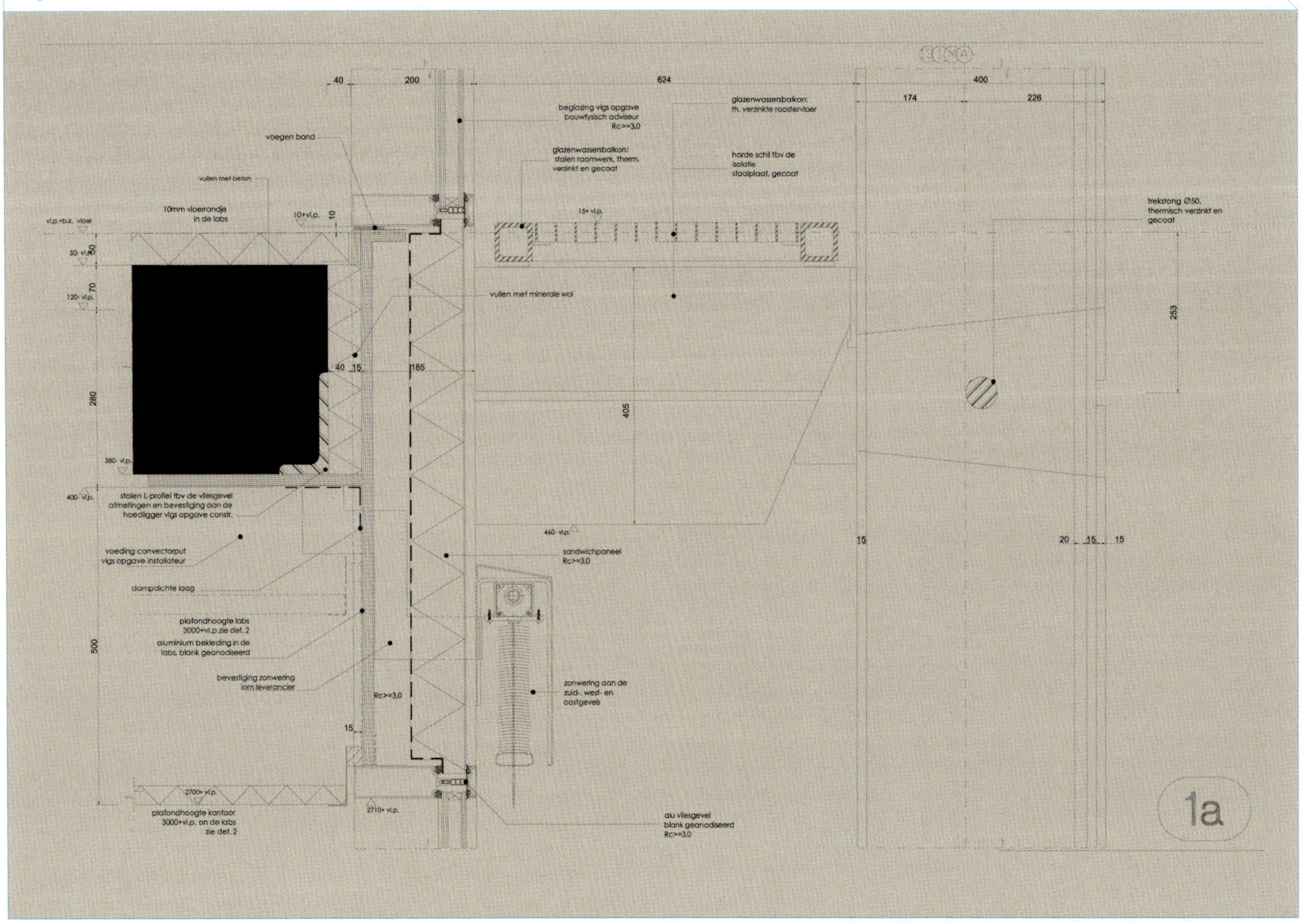

ATLAS BUILDING

VISIT ARCHITECTS WEBSITE

COLOR VISION OF FACADE

MUSAC IN LEÓN / Mansilla + Tuñón Arquitectos / www.mansilla-tunon.com

OFFICES 'LA DEFENSE' / UN Studio / www.unstudio.com

NETHERLANDS INSTITUTE FOR SOUND AND VISION / Neutelings Riedijk Architecten / www.neutelings-riedijk.com

METZO COLLEGE / EEA / www.eea-architects.com

RADICAL GEOMETRY FORM BY NORMAN FOSTER

FREE UNIVERSITY / Foster+Partners / www.fosterandpartners.com

PALACE OF PEACE / Foster+Partners / www.fosterandpartners.com

INTEGRATED REVITALIZATION OF URBAN TISSUE

BAIJURK-PROJECT / Archipelontwerpers / www.archipelontwerpers.nl

MAISON FOLIE / NOX / www.noxarch.com

NORTEL NETWORKS HEAD OFFICE / Bligh voller nield / www.bvn.com.au

NEIGHBOR OFFICE BUILDINGS IN THE ZILVERPARKKADE

OFFICE ZILVERPARKKADE / EEA / www.eea-architects.com

ZILVERPARKKADE D / Rene van Zuuk Architekten / www.renevanzuuk.nl

UNIQUE GLASS FACADE IN THE CITY

TOY MUSEUM / SCDA architects / www.scdaarchitects.com

CITRÖEN FLAGSHIP SHOWROOM / Manuelle Gautrand / www.manuelle-gautrand.com

FRIGORIFERI MILANESI AREA / 5+1AA / www.5piu1aa.com

HESSING COCKPIT IN ACOUSTIC BARRIER / ONL [Oosterhuis_Lénárd] / www.oosterhuis.nl

MONOLITHIC STRUCTURES

PAVILION NOIR / Architecte Rudy Ricciotti / rudy.ricciotti@wanadoo.fr

RICHARD E.LINDNER ATHLETICS CENTER / Bernard Tschumi Architects / www.tschumi.com

ATLAS BUILDING / Rafael Vinõly Architects / www.rvapc.com